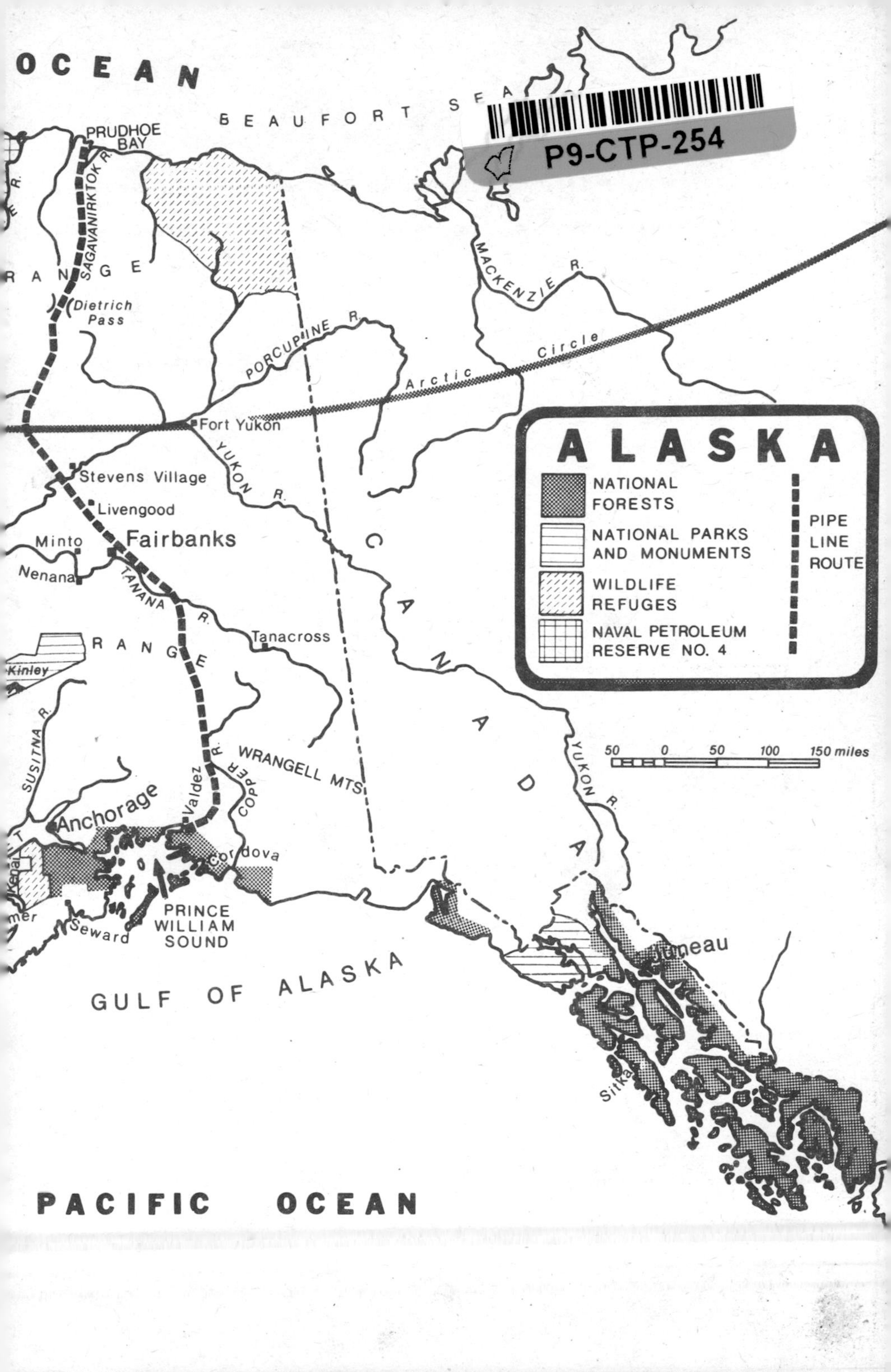
OCEAN
BEAUFORT SEA
PRUDHOE BAY
SAGAVANIRKTOK R.
RANGE
(Dietrich Pass
MACKENZIE R.
PORCUPINE R.
Arctic Circle
Fort Yukon
YUKON R.
Stevens Village
Livengood
Minto
Fairbanks
Nenana
TANANA R.
RANGE
Tanacross
CANADA
Kinley
SUSITNA R.
WRANGELL MTS.
Valdez
COPPER R.
Anchorage
Kenai
Cordova
Seward
PRINCE WILLIAM SOUND
GULF OF ALASKA
Juneau
Sitka
YUKON R.
ALASKA
NATIONAL FORESTS
NATIONAL PARKS AND MONUMENTS
WILDLIFE REFUGES
NAVAL PETROLEUM RESERVE NO. 4
PIPE LINE ROUTE
50 0 50 100 150 miles
PACIFIC OCEAN

THE

N

The Alaska Pipeline

THE POLITICS OF OIL AND NATIVE LAND CLAIMS

Mary Clay Berry

INDIANA UNIVERSITY PRESS
Bloomington & London

Published in Canada by Fitzhenry & Whiteside Limited, Don Mills, Ontario

Manufactured in the United States of America

Library of Congress Cataloging in Publication Data

Berry, Mary Clay.
The Alaska pipeline.

Bibliography
Includes index.
1. Petroleum industry and trade—Alaska. 2. Alaska pipeline. 3. Petroleum—Pipelines—Law and legislation —Alaska. 4.Land—Alaska. I. Title.
HD9567.A4B47 333.8′2 74-11716
ISBN 0-253-10064-X

Title page photo by John M. Berry

For my mother
Miriam Berle Clay

CONTENTS

The Alaska Pipeline

THE POLITICS OF OIL AND NATIVE LAND CLAIMS

1

SUMMER 1969

Late in the summer of 1969, political scientists, economists, conservationists, politicians, and oilmen met in Fairbanks to talk about Alaska's future. The state had just entered the league of the oil rich, the heady company of Iran, Texas, Kuwait, and the Arab sheikdoms. Only a little more than a year before, after years of expensive but futile exploratory drilling, the Atlantic-Richfield Company (ARCO) had found oil on the polar desert beside the Arctic Ocean. Estimates of the size of the petroleum reserves ARCO had tapped on the North Slope varied but even the most conservative, 9.6 billion barrels, indicated that the oil field there was a major one. The discovery generated the sort of excitement Alaska had not known since the turn-of-the-century gold rushes. Men, money, and machines poured into the state. There were even rumors the Mafia was coming. Those who held leases on the North Slope, purchased as speculation years ago, were offered millions of dollars for them. About 1,400 Outside corporations registered to do business in Alaska. (To Alaskans, anything beyond the state's borders is Outside.) Conglomerates began to assimilate Alaskan businesses, particularly construction firms with Arctic experience.

As the professors and politicians gathered in Fairbanks, public interest was focused on the September 1969 sale at which the state planned to offer oil and gas leases on more than 430,000 acres of the North Slope. Many Alaskans expected the oil industry to pay more than a billion dollars for the right to drill on land adjacent to Prudhoe Bay, where ARCO had struck oil. Alaska was mesmerized by that figure. One billion dollars was an almost inconceivable windfall in a state which is plagued with an inhospitable climate, high seasonal unemployment, persistent poverty, and a low tax base. Tension mounted and feelings ran high as September 10 approached. In this volatile atmosphere, the Twentieth Alaska Science Conference opened under the auspices of the state's university. Its purpose was to consider the changes oil would bring to Alaska.

The lead-off speakers were Robert Engler, a political scientist best known for his classic study of corporate power, *The Politics of Oil*, and Frank N. Ikard, then president of the American Petroleum Institute, the industry's major trade and lobbying organization. The local newspapers had billed the Engler-Ikard encounter much as they would a prize fight. And Engler did not disappoint them. He described in caustic terms "the penetrating powers of oil" that have often left public officials whose duty is overseeing the petroleum industry with an allegiance first to oil and then to the public.

"I thought I worked in the international petroleum industry," an Alaskan representative of British Petroleum Ltd. (BP), one of the giants with interests on the North Slope, remarked wryly. "But as he carried on with his list of original sins, I began to think I was in some Mafia-like organization." [1]

"I think it is utter folly to leave your judgments . . . to oilmen or to a vacuum of power," Engler warned Alaskans.[2]

Ikard and other oilmen skirted a direct encounter with their archcritic. As Engler himself documented in 1961, they presented themselves as so many concerned good citizens embarking on a joint venture with the people of the newly oil-rich state. In Fairbanks, oilmen talked about the way "the Alaska Oil Game" would be played, the stages of oil development, their effects on the land, the mystique of oil technology, and the sorts of laws and regulations that would bring the greatest benefits to everyone.

"I am convinced that a great and lasting oil industry here in Alaska can be developed, but only under a favorable price and tax structure," said Rollin Eckis, a top ARCO official.[3]

James H. Galloway, a vice-president of what was then the Humble Oil and Refining Company, joint owners with ARCO of one-tenth of the leased land on the North Slope including the discovery well, observed, ". . . the oil is going to be extracted and some of the country hitherto unmolested is going to be torn up in the process. Let's not fool ourselves. This activity is already far past the point of return." [4]

All the major conservation organizations were represented in Fairbanks. Conservationists were well aware that theirs was a cause whose hour had come. As filthy water and poisoned air afflicted most of the United States, its people became increasingly aware of the few places in the country where these conditions did not exist. One was Alaska.

Furthermore, in the summer of 1969, Congress was well on the way toward approving the National Environmental Policy Act (NEPA) of 1969, which was to become the conservationists' major tool for influencing public policy in the 1970s. This legislation, signed into law in early January 1970, required the federal government to make public a detailed list of all the adverse environmental effects of any project in which it was involved as well as the alternatives it had considered before approving any such project. Probably few Congressmen realized what they were doing when they voted for NEPA since it applied equally to the dredging of a local harbor and the construction of an anti-ballistic missile launching site.

The conservationists did not mince their words. Edgar Wayburn, vice-president of the California-based Sierra Club, observed, "Now that we have technology, we can grab more rapidly, more efficiently, and far more destructively than we could one hundred years ago—but we are still in there grabbing." [5]

Lynton K. Caldwell, a political scientist whose thinking helped to shape the 1969 law, said, "it is evident that in the minds of many Alaskans there is a belief in the continuing value of the practices of the traditional American frontier. Among these people there is a feeling that Alaska has a right to have its 'frontier,' that it has a right to the same kind of economic development, the same type of 'get rich quick' opportunities, the same philosophy of 'come and get it' that all the rest of the United States has had. And there is going to be 'moral' indignation when some federal bureaucrat, some Act of Congress, some conservation group says: 'No.' "[6]

Several speakers questioned the wisdom of the upcoming lease sale in view of the state's existing leasing and tax policies. One Alaskan economist, Gregg Erickson, of the University's Institute of Social, Economic and Government Research, urged the state to postpone the sale until the legislature could overhaul the leasing laws. He questioned whether the state would get anything approaching the true value of the resource under the bonus system Alaska used. He pointed out that the three North Slope giants, ARCO, Humble, and BP, had spent about five and a half million dollars for leases in the vicinity of Prudhoe Bay in the past, but what they got was valued later at more than two billion dollars. The idea behind the bonus system is to provide the state with its money immediately by requiring bonus bids at the time of the lease sale, rather than in dribs and drabs throughout the life of an oil field. It is also supposed to protect the state against loss should industry estimates of a field's worth prove too optimistic. The trouble with applying this theory to a field the size of the one at Prudhoe Bay, according to Erickson, is that even the international oil companies cannot pay in advance bonuses comparable to the value of the petroleum and natural gas there. However, to question the propriety of the Great Lease Sale less than twenty days before it was scheduled was tantamount to heresy for an Alaskan.

Most of the participants at the Fairbanks conference knew that oil had come to stay in Alaska and some expressed optimism that Alaska could control oil, not the other way around. Another

Alaskan economist, Arlon Tussing, suggested that the system by which the major oil companies have traditionally controlled prices and profits in their industry might break down in Alaska, resulting in "a complete reorganization of corporate and institutional relationships in an industry which has not been noted for its progressiveness or public spiritedness." [7]

And a British ecologist with long experience in the Arctic, Frank Frazier Darling, questioned the "unwarranted assumption" that Alaskans must choose between conservation and development. "Alaska is new enough and now on the way to being rich enough to build her economy and her physical structures and use her resources in ways that will not produce ultimate derelict landscapes," he said.[8]

The Arctic is littered with the junk of man's habitation there. "I used to think when I first came around the territory [of Alaska] as it was then," said Darling, "that men were arrogant in their prodigal creation of litter and filth as a token of their half-frightened attitude of conquering the wilderness." [9]

Trash endures in the Arctic. The oil drums which the United States Navy left behind in their 23-million-acre petroleum reserve near Point Barrow after nearly ten years of exploratory drilling in the 1940s and 1950s are only now beginning to rust.

Those oil drums in Naval Petroleum Reserve #4 were a muleta with which the conservationists taunted their bulls. During the conference a bull charged. United States Senator Ted Stevens, R-Alaska, addressing a luncheon at the end of the conference, said the barrels "enhanced" Pet Four.[10] He said a lot of other things at that luncheon so no one had an opportunity to find out how the barrels helped the reserve, but the next day he explained that what he meant was the Eskimos in Barrow could sell the oil drums for between $7 and $10 apiece. He did not say who would buy them.[11] Much later, the Senator asked Congress to appropriate $3 million so the Navy could remove the barrels.

Senator Stevens, who had intended to give an entirely different sort of speech that afternoon, threw away his text and delivered himself of his moral indignation. "I am up to here with people who tell us how to develop our country," he said.[12] Federal regulations governing construction of a proposed pipeline from the North Slope to an ice-free port in southern Alaska were "stupid, absolutely stupid." The purpose of these regulations was to minimize damage to the environment as the pipeline was built across the vast wilderness of interior Alaska, but Senator Stevens said they read as if Alaskans couldn't be trusted to manage their own land. A day later and unrepentant, the Senator, who had been a successful lawyer and state legislator before the Governor appointed him to serve out the term of a man who had died in office, defended himself. "Someone had to reflect what Alaskans are thinking," he said.[13]

The professors were angry but also smug. After all, the Senator's remarks were a perfect example of what Caldwell meant by the "frontier" frame of mind. Richard A. Cooley of the University of Washington, an authority on land use and abuse in Alaska, called the Senator "a man with Neanderthal mentality." Economist George W. Rogers, one of the men who had organized the conference, told his guests, "The luncheon speaker was not elected by the people of Alaska. If he speaks for anyone, he speaks for the man who appointed him to the Senate when Bob Bartlett died." [14] He was referring to former Governor Walter J. Hickel, by then Secretary of the Interior.

It is tempting to dismiss Senator Stevens as a "rapist" of the land, to borrow a phrase from the lexicon with which both sides assault one another in the Battle of Alaska, because he unleashed his temper at precisely the wrong moment from a public relations point of view. It is also facile to dismiss conservationists as "preservationists" who sought to "lock up" Alaska in one huge inviolable wildlife preserve. Indeed, it is easy to take from that 1969 conference whatever suits one's particular view of things. Almost every point of view was represented. But there was nothing simple about the conference itself. It addressed itself to change in the Arctic, always terrifyingly sudden, whether natural or man-made.

And the Arctic was changing, had already changed. The vice-president of Humble Oil was right. By 1969, it was too late to turn back.

And now a footnote to that conference, a footnote which should not be a footnote at all. The men who met in Fairbanks to discuss what to do with the land in Alaska all accepted the premise that it was their land to do something with. Conservationists pointed out that Alaska, most of which was under federal control, belonged to all the people of the United States. Oilmen professed to be working for the good of all people. And Alaskans like Senator Stevens talked about their country.

But one-sixth of Alaska's population had already rejected this premise—the Eskimo, Aleut, and Indian descendants of the aboriginal inhabitants of Alaska, whose unsettled claims to more than 90 percent of the state were soon to tie up future oil development in double bow knots. The Alaskan Natives (this catch-all term is capitalized as an expression of racial pride) said the land in question had belonged to their ancestors and, since it had never been sold or taken from them in battle, still belonged to them. The Natives did not insist upon title to all 375 million acres in Alaska, admittedly a political impossibility, just part of them. In 1969, most Natives were willing to settle for 40 million acres plus money for the rest of the land they claimed. They wanted those 40 million acres to include their traditional hunting and fishing grounds and the land around their meager villages, but their leaders also wanted land that would provide an economic base for their people in the future when the simple village life might no longer be possible. They wanted protection against the change which they also knew had come to Alaska.

Conferences on Alaska's future always had trouble coping with the Natives. After the September 10 lease sale, at which the state collected $900 million, there were several studies of what to do with the money. The state legislators invited the Brookings Institution to help them chart a course for the future. This produced a series of unusual and provocative forums on the state's problems and priorities, but Brookings did not touch the Native claims question, then a particularly volatile issue because of the vocal anti-Native backlash developing in the state. Meanwhile, Governor Keith H. Miller hired a team from Stanford University to advise his administration on what to do. The Stanford team eventually submitted a report which also ignored the Native question.

Actually, the Science Conference did better than most. The Natives were on the agenda; a number of anthropologists who had worked in the state spoke about them. But only one Native was invited to participate, an articulate Tlingit Indian leader named John Borbridge, Jr.

One thing Borbridge told the conference was that the Alaskan Native was accepted and tolerated as long as he was on the receiving end of a complex "industry" of assistance, the Bureau of Indian Affairs (BIA), and its Alaskan counterpart, the Alaska Native Service. "But what happens as the Alaskan Native assumes his rightful place as an equal partner in the economic, political, and other power structures of this state?" Borbridge wondered. "This may be a threat to a very comfortable power structure and I say power structure in its highest sense." [15]

Most of the Natives, who have a history of bemused tolerance toward the white man, found a grim amusement in their predicament. Three Native leaders, Borbridge among them, joked about asking the Ford Foundation for a grant to study white Alaskans but decided they should wait until they had more leisure in which to pursue their study.[16]

But not all the Natives were so philosophical. One young Eskimo from Barrow, Charlie Edwardsen, criticized the organizers of the conference for not inviting more Natives to participate. The organizers were dismayed, perhaps as much by his attire—unpressed khakis—as his manner. "After we worked so hard to bring all these important people together to focus on the problems of the natives," said one, "how could he?" [17]

The angry young man from Barrow eventually became something of a public nuisance, showing up at meetings to ask eminent conservationists where they stood on the Native claims (usually they were silent), picketing the lease sale with signs accusing the state of "economic genocide," calling the Alaska establishment and the oilmen "Pigs" in public, and twitting oilmen at Washington receptions. Throughout it all he wore his rumpled khakis, the sartorial badge of the Native Left.

In common with dissidents everywhere in the United States, Edwardsen did not get precisely what he said he wanted. While he insulted the establishment, the Native leaders (including Edwardsen) and other politicians came to grips with the realities which faced them and worked out a compromise that pleased no one totally but allowed all to get on with the process of developing Alaska. More than two years after the Fairbanks conference, the Alaska Natives got their 40 million acres, plus nearly $500 million in cash and a stake in the state's economic future in the form of a royalty on future mineral revenues. Despite its flaws, this settlement was an extraordinary agreement in view of the United States' traditional way of settling its Indian problems.

But it didn't settle everything. Three months later, the conservationists, the Natives, and the state were still arguing about what the Native claims settlement meant.

"That isn't much of a bill we wrote," remarked Senator Len B. Jordan, R-Idaho, sotto voce, during an informal Senate hearing on amendments to the 90-day-old legislation by which the Natives' claims had been settled.

"That was a lawsuit we wrote," whispered back his colleague, Senator Lee Metcalf, D-Mon., who had just asked three consecutive witnesses to explain a controversial section of the law. None of them could.

2
TWO ALASKAS

"We're trying to make a Fifth Avenue out of the tundra. . . ."

Walter J. Hickel, 1958

"They tell me Russians sold our land to the Government. There were no Russians on our land. There were no white people. White people never came there. They never saw it. I think Government buy stolen property maybe. Tough luck for Government. Can buy whole world that way."

Ewen Moses Laumoff, 1968

Alaska is the Great Land. It has abundant wildlife, a friendly nourishing sea, rapidly flowing rivers, sheltering forests. Alaska is also a harsh land, with a difficult climate, seasonal hunger, spectacular loneliness. It is a mistake to think of Alaska as one land; it is many. If Alaska were superimposed on the contiguous United States, the tip of its panhandle would be in Florida, the end of the Aleutians somewhere in southern California, and Point Barrow in northern Minnesota. It is a vast place.

There are rainy temperate forests in the panhandle, squeezed in among the spectacular snow-capped mountains of the Coast

Range. These rugged mountains follow the coast from the southern end of the panhandle around the Gulf of Alaska to the Kenai Peninsula, where they disappear into the sea only to re-emerge as Kodiak Island. Huge glaciers, the largest bigger than Rhode Island, spill into the sea. Fjords and rivers give occasional access to the land on the other side of the mountains. The towns are little more than a cluster of houses along a narrow beach.

North of the Coast Range is a second set of even higher peaks. The Alaska Range includes Mount McKinley, the highest mountain on the continent, which rises abruptly to 20,320 feet. This range crosses Alaska and extends far out into the Pacific as the Aleutian Archipelago, barren, rocky stepping stones to Asia.

Beyond the Alaska Range are the windswept plains of the Interior, where summer temperatures may rise as high as 100° and winter temperatures sometimes drop to 75° below zero. Life in the Interior is centered along its mighty rivers, the Yukon and the Kuskokwim, which end in a broad marshy delta on the Bering Sea, the nesting place for thousands of birds from every part of the world. The Interior is generally flat, punctuated with thousands of shallow lakes and covered with muskeg and tundra.

North of the plains, the northern end of the Rocky Mountains crosses Alaska for about 600 miles, ending in a sharp peninsula between the Arctic Ocean and the Chukchi Sea, which early American whalers called Point Hope. This is the Brooks Range, lower than other Alaskan mountains but no less rugged. Many rivers rise in the numerous glacial lakes here and the tundra nourishes herds of caribou and Dall sheep.

The North Slope of the Brooks Range, where oil was recently discovered, is the true Arctic. This is a land of low hills and plateaus, broad braided rivers and pebbled beaches. The soil is frozen throughout the year but covered with mosses, lichens, grasses, flowers, and one-inch-high willow trees, all of which burst into bloom in the continuous sun of the brief Arctic summer. While the weather here is more moderate than in the Interior, a fierce wind whips the land all year, making surival tenuous during the long dark winter.

Man has been living in Alaska for thousands of years. Anthropologists have found human traces in the Arctic dating from as early as 6000 B.C. The Copper River Valley in south central Alaska has probably been inhabited for 5,000 years.

The Native people vary as widely as the land on which they live, but they can be roughly divided into four groups: the Eskimo, who live along the northern and western coasts and to some extent inland; the Aleuts, also Eskimo, who inhabit the Aleutian Islands; the coastal Indians, primarily Tlingits and Haidas; and the Athapascan Indians in the Interior. For the most part these people had their own well-defined areas of "aboriginal use and occupancy," which were technically the basis for their twentieth-century claims against the federal government. But the boundaries between living areas were hazy and often overlapped. Traditionally, the different groups did not like one another much; the Eskimo word for Indian is "Ingalik," which means "having louse's eggs"—not the description of a friend.

The first white men to come to Alaska were Russians. Vitus Bering, a Dane in the service of the Czar, discovered Alaska in 1741. Several years later, Russian traders began to flock to the Aleutian Islands, where Bering first landed, in search of sea mammal pelts, particularly those of the sea otter. The Russians conquered and systematically slaughtered the hardy Aleuts, a seafaring people whose rough existence contrasted sharply with their dainty handiwork, grass baskets, and bird gut parkas. Those Aleuts who were not killed died of smallpox, measles, or influenza. When Bering first made contact with them, there were about 20,000 Aleuts. They were probably the most numerous Eskimo group in Alaska. But by 1910 there were less than 1,500, including persons of mixed blood.[18] Those Aleuts who survived the initial contact with the Russians usually became their slaves. In the 1780s many were taken to the previously uninhabited Pribilof Islands to run fur seal rookeries for their masters.

At first there were several Russian companies competing in Alaska, but by 1799 the Russian American Company was given monopoly rights there. In the beginning the company was run as a feudal fiefdom by Alexander Baranov, who pushed the boundaries of Russian Alaska north and east. He established a capital at Sitka although he lost the town once to the fierce Tlingit Indians who lived in the panhandle.

The Tlingits were a highly organized and sophisticated people, ethnically aligned to the coastal tribes who ranged from Washington State to Alaska. They had a distinct sense of territorial prerogative. Neither the Tlingits nor their neighbors, the Haidas, were ever conquered by the Russians. But since southeastern Alaska was rich in easily obtainable natural resources—furs, fish, and forests, the very things which had stimulated the development of the complex and highly materialistic Tlingit-Haida culture also doomed the area to early and continuous exploitation by white men. However, since the Russians traded with the coastal Indians rather than attempting to conquer them, they fared better than the Aleuts had at the hands of the newcomers, and their way of life did not change radically until after the purchase of Alaska by the United States in 1867.

Like the British East India Company, which was its model, the Russian American Company was really an arm of the Russian government. But as the Russians moved eastward, they encountered a similar organization, the Hudson's Bay Company of Great Britain. The Russians were threatened not only by the British on the east but also by American whalers who were beginning to use the Bering and Chukchi Seas on the west. The Russians, who needed more ships and soldiers to protect their holdings than the Czar could spare, increasingly feared that they would not be able to hold onto their Alaskan property.

Also, profits were dwindling. The market for seal and sea otter fur had shrunk and the Russians' technique for treating the pelts was inferior to that of other traders.

In 1867 Russian Alaska was sold to the United States for $7

million. The United States also bought out the Russian American Company's interests there for an additional $200,000, an afterthought which added little to the price of this real estate bargain. The Russians left little behind them—a few small Orthodox churches, the administrative buildings in Sitka, the seal rookeries on the Pribilofs, and the Russian surnames they had bestowed on the Native inhabitants as the result of intermarriage or conversion to Christianity—Kvasnikoff, Klashinoff, Lekanof, Oskolkoff.

Just as the Russians altered Aleut and coastal Indian life through their fur-trading activities, the Americans were to change Eskimo life through whaling. The United States had become seriously interested in Alaska about twenty years earlier. In 1848 commercial whaling began in western Alaska. A few years later, whalers were regularly traveling along the Arctic Coast as well. The whalers came for baleen, a flexible horny substance found in the mouths of certain whales. Bowhead whales use baleen to strain food from sea water. White men used it for corset stays. The value of baleen rose from 32 cents a pound in 1850 to $4.90 a pound in 1905. Other parts of whales also had commercial uses, particularly the oil, but when a substitute was found for baleen in women's corsets, whaling in northern Alaska ended.

When the whalers arrived in the Arctic, they found two groups of Eskimos—the Tariamiut or Sea People and the Nuunamiut or Land People. The Tariamiut generally lived in well-defined villages from which they hunted sea mammals. The Nuuniamiut were nomadic since they were dependent upon the migratory caribou for food. Despite their isolation, both had frequent and well-established contacts with one another. An elaborate network of trade routes, generally running from west to east, enabled them to benefit from one another's cultures.

This important trading system continued after the arrival of the first white men and enabled Alaskan Eskimos to obtain English guns and ammunition from the Hudson's Bay outpost on the Mackenzie River. But whaling put an end to the system since the whaleboats could distribute the goods the Eskimos wanted more efficiently than the Native traders could. Abandonment of the historic trade routes forced most of the Nuuniamiut to move to the coast. In fact, the residents of one small village in Anaktuvuk Pass in the Brooks Range are probably the only Land People who remain in their traditional territory.[19]

The Eskimos' lifestyle was dictated by their harsh environment, whether they were maritime people or hunters of caribou. Quite simply, they went where the food was. Today, airplanes and the *North Star*, the Bureau of Indian Affairs ship which visits the coastal villages once a year, bring in flour, sugar, coffee, tobacco, guns and ammunition, building materials, fuel oil, and many other items without which the modern Eskimo cannot live as he wishes. But the villagers are still dependent upon hunting and fishing for much of their food. Although the people now live in permanent villages, they may spend weeks or months in search of food, often taking their families with them to live in tents. Indeed, a successful village Eskimo today is a person who is able to take advantage of both cultures—and this is no easy trick.

Just as the Russians had slaughtered the seals and sea otters, the Americans systematically exterminated the whales. After 1915, Eskimos along the northern coast who had depended upon whales for food faced some lean years. And although the whalers generally treated the Eskimos more humanely than the Russian traders had treated the Aleuts, contact with them was still disastrous for the Native people. The white man's diseases wiped out whole villages. Whiskey became the scourge of others. (Even today its sale is banned by many village councils.)

Historically, the broad delta of the Kuskokwim and Yukon Rivers was a place where different peoples met. Its earliest inhabitants were probably Eskimo, but by the time the Russian traders came there in the 1840s, Eskimos and Indians were living along both rivers. Today the Native village of Bethel is located in this delta area.

Little is known about the Indian settlements in the Interior. Some villages along the Yukon had contact with Russian traders from the delta, but for most, the first real experience with white men came in the 1890s with the gold rushes. Further east was the territory of the Kutchin, a group of Athapascan tribes which lived in both Alaska and Canada. Their first contact with white men was with the British in 1847, when the Hudson's Bay Company set up a trading post at the junction of the Yukon and Porcupine Rivers. Other tribes lived along the Tanana, a major southern tributary of the Yukon.

In south central Alaska, white men had greater impact on Native life than in the Interior. Around Bristol Bay, just north of the Aleutians, Russians traded for the first sixty years of the nineteenth century. South and east of the Aleutians, Kodiak Island, a major stronghold of the Russian American Company, supported a relatively large Eskimo community. The Koniags had probably lived there and on the adjacent mainland for 3,000 years. But a Russian visitor in 1805 told of villages inhabited by nothing but starving children. The Russians had taken away the hunters.

Further east along the mainland coast was Indian territory. An Athapascan tribe, the Tananina Indians, lived where Anchorage, now the largest city in Alaska, is located. Here the Indians resisted the Russians until 1838, when, greatly weakened by smallpox, they were conquered by them. By the 1960s only two identifiable Native villages remained, Tyonek and Eklutna.

More Athapascan Indians lived around the headwaters of the Copper River. To the south, along the Gulf of Alaska, was an area of racial mixing. At the west end of the Gulf coast lived the Chugach, the most southern of the Eskimo tribes. A small group of Indians called the Eyaks lived in the middle around the Copper River delta. The Tlingit Indians used the eastern shores of the Gulf although they had no permanent villages west of Yakutat on the panhandle.

The United States acquired Alaska in 1867 but did little to assimilate it. The Treaty of Cession gave it most of the land in modern Alaska plus the Russian American Company's trading rights. However, modern Alaska Natives believed that what the

United States got was not title to the 375 million acres but administrative sovereignty over them. They based their interpretation upon Article Three of that treaty, which deals with the rights of the people living in Alaska, some of whom were Russian citizens:

> The inhabitants of the ceded territory, according to their choice, reserving their natural allegiance, may return to Russia within three years; but if they should prefer to remain in the ceded territory, they with the exception of the uncivilized tribes, shall be admitted to the enjoyment of all the rights, advantages, and immunities of citizens of the United States and shall be maintained and protected in the free enjoyment of their liberty, property, and religion. The uncivilized tribes will be subject to such laws and regulations as the United States may, from time to time, adopt in regard to aboriginal tribes of that country.

After its purchase, Alaska was governed for seventeen years by the military, which, for all practical purposes, meant no government at all. Thus by 1880 less than three hundred white men lived in Alaska and all but thirty of them lived in Sitka.[20] However, United States business interests had taken the place of the Russian American Company. A California group, the Alaska Commerical Company, was given a twenty-year franchise on the seal rookeries in the Pribilofs, despite the fact that they had offered the lowest of thirteen bids for this privilege. (The company, it turned out, had a friend in the United States Senate.)

During this period, the Natives did not exist officially. United States Indian policy was stated in the Northwest Ordinance of 1787: "The utmost good faith shall always be observed towards the Indians; their lands and property shall never be taken from them without their consent; and in their property rights and liberty they never shall be invaded or disturbed, unless in just and lawful wars authorized by Congress." However, it was a policy generally observed in the breach. By the time Alaska was purchased, the government was busy rounding up what remained of the tribes in the United States proper and putting them on reservations which were generally located on land no white man wanted. In 1871 Congress forbade any further treaties between the federal government and Indian tribes, thus depriving the Natives of the opportunity for any further land settlement.

The Organic Act of 1884 raised Alaska's status a notch, from a customs district to a land district. Among other things, it extended United States mining laws, principally the Mineral Location Act of 1872, to Alaska. This became important when gold was discovered near Juneau only a few years later.

The Organic Act of 1884 did take note of the unusual legal position of the Alaska Natives, providing "that the Indians or other persons in said district shall not be disturbed in the possession of any lands actually in their use or occupation or now claimed by them but the terms under which such persons may acquire title to such lands is reserved for future legislation by Congress. . . ." This became the legal basis for the fact that the Alaska Natives' claims were settled legislatively rather than judicially.

The legislators who drafted the Organic Act knew very little about either Alaska or the Natives, who outnumbered the white men there. They set up a special commission to report "upon the conditions of the Indians residing in said territory, what lands, if any, should be reserved for their use, what provisions shall be made for their education, what rights by occupation of settlers should be recognized." The following year the commission recommended that the general land laws of the United States be extended to Alaska. The commission also recommended that, as the Natives claimed "only the land on which their homes are built and some garden patches near their villages," bona fide settlers should be encouraged to come to Alaska to "open and develop its resources." Accordingly, Congress began to extend other land laws to the new district. Among these were the homesteading laws, which proved virtually inoperable there since almost none of Alaska had been surveyed. Furthermore, homesteading had been designed with farming in mind; however, most of Alaska was not suitable for farming. The Natives could not even acquire land in this fashion because they were not citizens.

Despite unworkable land laws, white men began to exploit in earnest Alaska's abundant natural resources in the 1880s and 1890s. The Alaska Commercial Company was flourishing in western Alaska. A salmon-canning industry began in southeastern Alaska in 1878 and within a decade spread westward along the coast. From the 1880s on, gold and copper were the mainstays of the Alaskan economy. The center of gold production was Fairbanks although there were deposits elsewhere in the district. The center of copper production was the Copper River Valley, where the Kennecott Mines were located.

Throughout this period of exploitation at the turn of the century, few people paid much attention to the Natives. Although they comprised a majority of the district's population, officially they scarcely existed. In contrast to the general thrust of United States Indian policy at the time toward assimilation of the Indian into the mainstream of American life, the Alaska Natives had no way to be assimilated. In 1887 Congress had passed the Indian Allotment Act, which was designed to encourage Indians to settle down on small farms and live like white men. The idea was to split existing reservations up into individual lots on the theory that ownership of property would make the Indians more palatable to white society. However, the new law did not apply in Alaska.

Meanwhile, southeastern Alaska was becoming settled, and as white men increasingly encroached upon their traditional hunting and fishing grounds, the Tlingit and Haida Indians protested. They wrote the Secretary of the Interior about this, but he replied, "I have to inform you that these matters all lie outside the control of this department and would be proper subjects for consideration of Congress." [21] The Indians then asked for a reservation, but since reservations were out of fashion, their request was ignored.

Various persons, including a commissioner of the Alaskan Land Office, urged that public land laws be extended to the Natives, but no one listened to them. However, the Supreme Court ruled in a 1905 Alaskan case, *Berrigan* v. *United States*, that the United States had an obligation to protect the property rights of its Indian wards. The following year, Congress passed the Alaskan equivalent of the Indian Allotment Act. This law allowed Eskimos and Indians (but not Aleuts) to apply for 160-acre homesteads on nonmineral land chosen from vacant and unappropriated parts of the public domain, that is, from unreserved federal land. But homesteading was no more practical for Natives than for whites, and perhaps even less so. Fifty-four years later, only eighty allotments had been issued, most of them in the southeast.

Even as the private interests were taking more and more land from Alaska, areas of the district were being set aside to protect them from exploitation. An elitist conservation movement sprang up in the United States about 1900, based primarily on the realization that the nation's natural resources were not inexhaustible. One government official who understood the need to conserve natural resources was President Theodore Roosevelt's Chief Forester, Gifford Pinchot. Pinchot believed in "the greatest good for the greatest number in the long run" and toward this ambiguous end he created the first of the nation's national forests in Alaska. The exact definition of the greatest good for the greatest number in terms of the use of these forests is still a point of controversy.

From the turn of the century on, the federal government took millions of acres in Alaska out of the public domain for various conservation purposes—national forests, national parks, wildlife refuges, petroleum reserves. Because certain restrictive regulations, differing from reserve to reserve, impeded economic development, many white Alaskans resented these withdrawals. They also argued, with a good deal of validity in many instances, that reserves were unrealistically administered and that frequently the administrative agencies had no clear plans for what to do with the land.

The first major reservation in Alaska was the 16-million-acre Tongass National Forest in the southeast, which was created in 1902 and enlarged seven years later. In withdrawing the land for this forest, the federal government paid little attention to existing settlements, whether white or Native. Much later, the Court of Claims ruled that the United States owed the Tlingit and Haida Indians $7.5 million for lands taken from them without compensation when the forest was set aside.

In 1906, Alaskan coal fields were withdrawn from entry so that no coal could be mined. In 1907 more than four and a half million acres around Prince William Sound in south central Alaska were set aside for the Chugach National Forest. The following year, Congress authorized the President to make withdrawals by Executive Order without Congressional approval. Although the Pickett Act applied to the entire United States, it was particularly controversial in Alaska. In 1916 the first Alaskan national park, Mount McKinley, was set aside, and a few years later a huge national monument was created at Katmai. The next big bite out of the public domain came in 1923 when President Harding created the 23-million-acre Naval Petroleum Reserve #4 in the Arctic.

In 1912 Alaska became a Territory and was given both a legislature and a nonvoting representative in Congress. One of the early acts of the territorial legislature was to enfranchise the Natives although no one was certain that the legislature had the authority to do this. It was not until 1924, when Congress declared all noncitizen Indians born within the territorial limits of the United States to be citizens, that Alaska Natives acquired citizenship.

Meanwhile, during this helter-skelter treatment of the Natives and their land, American Indian policy was experiencing one of its characteristic swings of the pendulum. The publication in 1928 of the Meriam Report marked the beginning of a policy which stressed the uniqueness of Indian culture and sought to protect and nurture it. (This remained official Indian policy until the 1950s, when the desire to assimilate the Indian into the mainstream again prevailed.) The Meriam Report led to the passage of the 1934 Indian Reorganization Act (IRA). This legislation provided new forms of economic assistance to Indian communities, enabling them to set up their own industries. It also provided for corporate ownership of tribal property and gave the Secretary of the Interior the right to enlarge existing Indian reservations and create new ones. At first only some of these provisions applied to Alaska, but in 1936 all of them were extended to the Natives in the territory.

In 1943 Secretary of the Interior Harold L. Ickes announced the creation of several Native reservations, pending approval by 30 percent of the Native residents in a special election. One was the 1.4-million-acre Venetie Reservation in the northeastern corner of the Territory, established for two villages on the Chandalar River, a tributary of the Yukon. There were 217 people living there, all but two of whom were Natives. The Secretary's action greatly alarmed white Alaskans, who feared that soon between one-third and one-half the Territory would be closed to them. Many whites thought Ickes' men had pressured the Natives into voting to create this and other reservations, and indeed the courts later found one reservation, at Hydaburg in the southeast, to have been improperly created for this reason.

However, the whites' fears were largely unjustified, for only six IRA reservations were actually created. Four others were proposed but were voted down by the Native residents, and the one at Hydaburg was eventually disbanded. The Bureau of Indian Affairs later proposed another eleven reservations, all in northwestern Alaska, which would have set aside 2.2 million acres for the use of about 2,000 people, but the Natives there never had an opportunity to vote on them. Eleven villages petitioned the BIA for reservations, but it took no action on their petitions. In the 1950s about ninety villages asked to have reservations created but by then "termination" was the official Indian policy and reservations were out of style again.

More important than the six reservations actually created in the 1940s is the present-day ill will which has its roots in that troubled period. Until 1939, although the Natives were a majority of Alaska's population, they were no threat to the few white residents and the two groups lived side by side in what appeared to be an amicable coexistence. As in many colonial stiuations, there was a good deal of intermarriage and the distinctions between the races sometimes grew fuzzy. After 1939, however, the white population of Alaska grew steadily. This expansion brought white interests into conflict with Native interests, and as the whites prospered the Natives got poorer and poorer. There was considerable racial discrimination, particularly after World War II began. As a visiting journalist observed in 1942, "while the white Alaskans are usually democratic among themselves, their attitude toward the Indians is deplorable—almost reminiscent of the southerners' attitude toward the Negros—and they are only too glad to have the federal government assume the burden of the natives' needs." [22]

At the same time, industries showing interest in Alaska were frightened by the spectre of unsettled Native land claims. In 1935 the Tlingits and the Haidas had persuaded Congress to allow them to seek compensation from the Court of Claims for lands taken from them when the Tongass Forest was formed. There was continual speculation on the validity of other Indian claims in the southeast. The Natives were at last a threat to development by whites.

World War II thrust Alaska into the twentieth century. The territory was important to the military as a jumping off point for Asia. The Aleutian Islands stretch invitingly across the Pacific toward Japan. In fact, early in the war the Japanese captured two of them—Attu and Kiska. At the start of the war, the United States military in Alaska consisted chiefly of a handful of men stationed at the antique Chilkoot Barracks in the panhandle. But soon army bases, airfields, and naval stations sprang up everywhere, and Congress authorized construction of a highway between the "lower forty-eight" and Alaska, which was built soon after the bombing of Pearl Harbor.

Late in 1939 there were slightly more than 70,000 people in the Territory. By 1950 the permanent population was about 138,000 and growing rapidly. The economy boomed.

After the war, construction continued as the cold war and its increasingly sophisticated weapons, particularly the long-range bomber, made Alaska even more important as a military base. First huge air force bases and then the Distant Early Warning (DEW) Line System were constructed. The second construction boom lasted well into the early 1950s but gradually defense spending in Alaska leveled off. In 1954 the Defense Department spent $416.9 million in the Territory. By 1966 the Defense Department budget for Alaska was down to $315.3 million.

There was an increase in civilian construction as well as military. During the war new industries sprang up to meet the needs of the many civilians who came to the Territory for high-paying construction jobs. One such need was housing. The civilian construction boom continued as people flooded into the Territory after the war. Many were veterans who had been stationed in Alaska during the war, had liked it, and had returned afterward with their families to live. Congress passed legislation to make homesteading easier for these men but, as previously, homesteading did not work well. Despite some spectacular and well-publicized success stories, most new arrivals found they needed money to make money. While jobs were plentiful in the early 1950s, the high demand for housing made it hard to find, and the cost of living was much higher than on the West Coast. The Armed Services urged personnel stationed in Alaska after World War II to bring their dependents with them, adding to the demand for housing and services.

By the mid-1950s, military interest in Alaska was waning. However, the Territory was beginning to reestablish an economy based upon natural resources. Oil was discovered in Kenai in 1957, the first oil produced in Alaska since the insignificant amount in the 1920s and 1930s at Katalla. Also, several companies indicated an interest in building pulp mills in southeastern Alaska. The fisheries, depleted by the lack of good conservation practices, were in bad shape, but many Alaskans believed that if the Territory could wrest control of them from the Seattle canneries, they would thrive again in a few years.

During this period of economic growth, the Natives were growing increasingly aware of their rights and asked repeatedly for the protection of reservations. Their petitions were ignored. In 1946 Congress had created the Indian Claims Commission and thus had given Indians who were unable to sue in the Court of Claims a forum for consideration of their grievances. However, few Natives learned about the Claims Commission before 1951, the deadline for filing claims. It is ironic that the Native claims, which were to create such problems for Alaska twenty years later, could probably have been settled in the early 1950s by the creation of numerous small reservations.

The Natives' growing uneasiness coincided with the white man's push for statehood for Alaska. While most proponents of statehood were aware of the Native land claims, few seem to have understood them and most thought that any attempt to settle them at the time of statehood would merely postpone everything. So, almost to a man, they disclaimed any responsibility for them. As one witness told a Congressional committee considering statehood, "The Indians, with their aboriginal rights, are a federal problem. We have no control over it and we cannot dispose of it and we have nothing to say about it. Whatever happens to Alaska it will still be a federal problem." [23]

No one wanted to talk about the claims. This issue was a highly emotional Pandora's box: to open it would let out bigotry and greed and fears that were inappropriate in a group of people petitioning for admission to the democratic United States of America.

The two strongest arguments for statehood were exploitation by "Outside" interests and neglect by the federal government. The activities of the miners and the Seattle "fish trust" won many friends for statehood. Federal neglect was harder to explain, but statehood proponents tried to show that Alaska's isolation led to unrealistic land use policies.

The chairman of the Alaska Statehood Committee, Robert B. Atwood, who was then and still is the editor and publisher of

Alaska's largest newspaper, the *Anchorage Daily Times*, told a House subcommittee in 1953:

> The effort to lock up the remaining resources was so thorough and intense that the federal government forgot provision for unlocking them. Alaska was closed to development under the guise of conservation. It remains for the most part today. The true meaning of conservation is controlled use and wise management. But in Alaska conservation, as it is and has been practiced, means paralysis. . . . this so-called conservation is actually a waste of the worst sort.[24]

Atwood's remarks are not unlike those of his fellow Alaskan, Walter J. Hickel, Secretary of the Interior from 1969 to 1971. A few weeks after his appointment to the Cabinet was announced, Hickel held a press conference at which he expressed opposition to "conservation for conservation's sake," in the best Alaskan tradition. In January 1969, at confirmation hearings before the Senate Interior and Insular Affairs Committee, the Alaskan was asked to explain what he meant. "Now, as an overall policy, I am for conservation, for the wide utilization and conservation of our resources," said Hickel. "I would say that when I made that statement, I was thinking of areas in my own country where there are millions upon millions, and I am not stretching this, board feet of timber that are just rotting because they have never been harvested. I do not think that is a wise use of natural resources." [25]

Proponents of statehood based their optimism about Alaska's future on projections of the four major growth industries: petroleum, wood products, fisheries, and tourism. They usually added some other factors like Alaska's untapped potential for hydroelectric power. They generally assumed that Alaska's economy would grow automatically, given favorable political institutions. As Alaskan economist George W. Rogers has described it, "If Marx had produced his theory of economic determinism by standing Hegel on his head, why not create a theory of political determinism by standing Marx on his head?" [26]

It was in those euphoric days of 1958 when the statehood battle was nearly won that Hickel, a budding entrepreneur with his public service still far ahead of him, made a remark which typifies the unrealistic attitude of many Alaskans to statehood's possibilities. Hickel told a *Time* reporter, "We're trying to make a Fifth Avenue out of the tundra. . . ."

Of the many problems associated with statehood three are especially significant in any discussion of Alaska in the 1970s: the question of how the new state was going to survive economically, the question of what to do with the vast federal lands there, and the problem of the Native land claims.

Congress devoted time and thought to giving the new state an adequate economic base. It finally settled upon a land grant of unprecedented size to provide revenues. Most western states in which sizable portions of the land were still not privately owned at the time they entered the Union received a land grant of two sections, or 1,280 acres, out of each township in the public domain for the support of schools. (A township contains thirty-six sections.) Under such a formula, Alaska would have gotten about 21 million acres of public domain lands to administer for revenue purposes. Instead the state got 102 million acres, more than the total land grant to all other western states combined. This amounts to roughly one-third of the total acreage of Alaska.

Members of the House and Senate Interior Committees which drafted the statehood bill, most of them representatives of land grant states, gave a lot of thought to this land formula. They abandoned the time-honored precedent of numbered sections for several reasons. First, since Alaska was unsurveyed for the most part, there were few townships. To survey the Territory before statehood was out of the question because it would take too long. Secondly, in Alaska, where the land is of varying value, "in-place" grants could have resulted in the state's getting title mostly to land with little foreseeable economic value. Finally, from a land management point of view, having small tracts of state land isolated from one another is unwieldy. Instead Alaska was to choose its revenue land anywhere in the public domain, but in reasonably compact tracts.

Statehood gave the state government the right to choose the following amounts of land, some within ten years, others within twenty-five: 102,350,000 acres from the unappropriated public domain for general purposes; 400,000 acres from the national forests in southeastern Alaska for community expansion; and 400,000 acres from the public domain for the same purpose. The legislation also confirmed earlier grants of federal land to the Territory, about 1.1 million acres in all.

Congress departed even further from precedent to make sure that Alaska would be economically viable. Before 1927, no state was allowed to choose so-called mineral lands for itself, that is, lands where the primary use would be mineral extraction. If any of its numbered sections were mineral in character, a state had to take an alternative section. After 1927, states which had not completed their land selections could take mineral lands, but the mineral rights thus transferred were inalienable, i.e., the minerals could be leased but not sold. Alaska was the chief beneficiary of this 1927 legislation.

In addition, Alaska was given a larger share of the revenues from mineral leases on public domain lands within its boundaries than any other state. Other western states receive 37.5 percent of these revenues directly. Ten percent goes to the federal government. The rest—52.5 percent—is paid into the federal reclamation fund for irrigation and land reclamation projects. Since Alaska is not a reclamation state, Congress provided that it should receive the full 90 percent of mineral revenues, including all receipts from rentals, royalties, and bonuses on leases on federal land. Alaska is the only state which has this large and continuing source of revenue from federal lands.

Congress' generosity gave the new state government a lot of land to manage. The Alaskan Constitution (drafted before statehood, partly to prove that Alaskans were capable of governing themselves), was one of the first state constitutions to contain a statement of general land use policy. Article VIII states: "It is the policy of the state to encourage the settlement of its land and the development of its resources by making them available for maximum use consistent with the public interest." The constitution empowers the legislature to "provide for the utilization, development, and conservation of all natural resources belonging to the state, including land and waters, for the maximum benefit of its people." The first legislature drafted a complex but flexible set of guidelines for land management. Alaska's fish, wildlife, and waters are reserved for the people of the state "for common use." And the state government is directed to manage and develop "the replenishable natural resources, the forests, wildlife, fish, and grasslands," on the sustained yield principle, "subject to preferences among beneficial users." The state cannot sell, grant, or deed its rights to these resources; it can lease them.

The question of what to do with federal lands in Alaska brought into conflict half a dozen powerful departments and agencies in Washington. Resolving it meant working out a delicate compromise, but it also raised still another question: just what was the national public interest in Alaskan land and how should it be protected?

In the end, Congress resisted Alaskans' pleas to trim federal holdings in the state and kept the national parks, wildlife refuges, forests, and other reserves. Whether it did so as a matter of conscious policy or because it was impossible to persuade the various federal agencies to give up any of their land is not so important in the long run as the fact that Congress did hold onto the land, thus protecting the broader national interest.

In 1958, when Congress approved statehood, there were 92.4 million acres of Alaskan land in federal reserves of one sort or another. There were more than 20 million acres of national forests, a 23-million-acre naval petroleum reserve, more than 27 million acres in power reserves (including a second Arctic petroleum reserve), 7.8 million acres of wildlife refuges, and 6.9 million acres of national parks and monuments. The federal government was trustee for more than 4 million acres of Indian reservations. The public domain consisted of 271.8 million acres, from which the new state was to get nearly 103 million. Only 700,000 acres had been patented to private individuals. There were unperfected entries on another 600,000. Thus Congress had kept roughly one-fourth of Alaska as federal land.

The other problem was the Native claims. When the Senate Interior Committee was in Ketchikan in 1953 for hearings on statehood, one of the witnesses, Eugene Wacker, told the Senators, "By the way, while I am here, there is a chief who asked me to bring this before you gentlemen, to try to settle the Indian aboriginal rights, Chief Kaihan."

"Did he want it settled in the next hour?" asked Senator Henry M. Jackson, D-Wash., archly.

"No," said Wacker, "he has been waiting for eighty-some-odd years."

"I do not want to be unduly pessimistic," interjected Senator Clinton B. Anderson, D-N. Mex., "but it may still run that long again."

Wacker did not persist. He simply said, "He asked me to bring it up."[27] So much for Chief Kaihan.

Although proponents of statehood said that settlement of the claims was a federal problem and not an Alaskan one, the Natives themselves were largely silent. Most of the village people were unaware of the way in which statehood might affect them. At that time there was only one statewide Native organization, the Alaska Native Brotherhood and Sisterhood (ANB), and even it was statewide in name only, for most of its membership came from the southeast. The main concern of informed Natives and of national Indian organizations was that Congress not extinguish the claims when it created the state of Alaska. The Organic Act of 1884 had reserved determination on the claims "for future legislation by Congress," and the Natives were fearful that Congress might use this mandate to simply wipe out their claims. They wanted to preserve whatever rights they might have, although at that time it was unclear what they were.

The members of Congress who drafted the statehood bill were also concerned about what their actions might mean vis-à-vis the claims. On the one hand, they wanted to leave all existing claims undisturbed. On the other, they did not want to give the claims any legal validity they did not already have.

It is clear from their conversations, recorded in an unusual set of public mark-up sessions in 1954, that many Senators did not think the claims were valid. Others used the unsettled claims as one more argument against statehood. And at least one member, Senator Anderson (perhaps haunted by the spectre of Mexican-American claims in his own state of New Mexico), hoped that including specific language in the Statehood Act would force adjudication of the claims in the near future, thus opening up Alaska to economic development.

When the Committee met in 1954 to iron out the details of one statehood bill which subsequently died without coming to a vote in the House of Representatives, the Justice Department advised the Senators not to mention the Natives' rights at all even if only to disclaim them. Senator Guy Cordon, R-Ore., who was running the meeting, agreed with the Justice Department. And Senator Jackson, who was to become the chief arbiter of the claims question ten years later, agreed with Senator Cordon. Senator Jackson raised the further point that to include a disclaimer might be to "recognize the existence of a right on the part of the natives." [28] He argued that if no disclaimer were included, the legislative record would be sufficient to show that Congress was merely maintaining the status quo until the courts could act. At the time, in fact, three Alaskan cases were pending before the Court of Claims, one of them the Tlingit-Haida case, in which the Indians later won a compensatory payment from the United States. Alaskan Delegate E. L. Bartlett argued that the absence of a disclaimer of Native rights would lead to the interpretation that Congress did not think there were any aboriginal rights involved and would be certain to anger Indian organizations across the country.

"The feeling of the representative of the Department of Justice was that if this language was left in the bill the courts might construe this section as giving and recognizing a right which the courts refused to recognize in their decisions to date," replied Senator Jackson. He suggested the Committee spell out precisely what it meant by the disclaimer clause in its report.

In all this discussion, the Senators' primary interest was not the Natives' well-being. Instead they wanted to insure that companies that came to Alaska to develop it would be able to get clear title to the land. "Unless we can find some method by which we can say to them, 'So far as you are concerned, the title is in the state and it is a good title,' then we cannot expect the kind of investment that is absolutely essential if the state is to develop," warned Senator Cordon.[29]

Four years after this frank discussion, Congress finally approved Section 4 of the Statehood Act of 1958, which includes a disclaimer after all:

> As a compact with the United States, said state and its people do agree and declare that they forever disclaim all right and title to any lands or other property not granted or confirmed to the state or its political subdivisions by or under the authority of this act, the right or title to which is held by the United States or is subject to disposition by the United States and to any lands or other property (including fishing rights), the right or title to which may be held by Indians, Eskimos, or Aleuts . . . or is held by the United States in trust for said natives; that all such lands or other property (including fishing rights) the right or title to which may be held by the United States in trust for said natives, shall be and remain under the absolute jurisdiction and control of the United States until disposed of under its authority, except to such extent as the Congress has prescribed or may hereafter prescribe, and except when held by individual natives in fee without restrictions on alienation.[30]

While legal arguments were presented during the land claims settlement that Section 4 invalidated the Natives' claims, Congress decided that this section merely reaffirmed its right to settle them.

On August 26, 1958, the majority of the voters in Alaska accepted the Statehood Act of 1958. The first legislature convened in Juneau in 1959 and drafted a variety of land laws. It created a Department of Natural Resources which, through its Division of Lands, would choose, manage, and dispose of the state's 102 million acres from the public domain. Although the legislators laid down some guidelines for the Division of Lands to follow, it was up to this agency to determine the state's objectives in its land selection program and to decide which among the possible uses of the land was in the public interest.

Slowly, slowly, slowly, for no one knew what most of the land in Alaska contained, the Division of Lands began to choose a tract here and a tract there. But very soon something happened which foretold the shape of things to come. In 1961, the BIA filed protests to state selections on behalf of four Native villages. These villages claimed about 5.8 million acres near Fairbanks. The state had already filed for patents to 1.7 million of them.

3

THE NATIVES AWAKE

"We are not a chess game. We are human beings and right now a very upset and disturbed people."

Andrew Issac,
Chief of Tanacross

In February 1963, the Alaska Conservation Society called a meeting in Fairbanks to discuss a summer recreation area the state had proposed about fifty miles from that city. The Minto Lakes were well known to local sportsmen for their abundant moose and small game. The state wanted the land and planned to build a road into the lakes from Fairbanks. In fact, the legislature had appropriated funds for the road two years earlier but the state could not get the necessary federal funds to build it until the Bureau of Land Management (BLM) approved Alaska's application for patent to this land as part of its 102-million-acre birthright. The meeting, attended by local sportsmen and conservationists, went along smoothly until the Society produced a surprise guest.

He was Richard Frank, a 36-year-old Athapascan Indian with a sixth-grade education. Frank was chief of the Minto Flats people, whose permanent village was Minto on the Tanana River below

Fairbanks. Minto was a river town. Its people hunted and fished there and found seasonal employment with the barge companies that ply the river's waters. But most of the 150-odd persons living in Minto also ranged over more than a million acres in search of food, including the lakes where the state planned its recreation area. Frank told the sportsmen that the lakes belonged to his people, that this was their traditional hunting ground, and that without the use of the lakes, his children and his people's children would go hungry. There was absolute silence while Frank spoke.

"Nothing is so sorrowful as for a hunter, empty handed, to be greeted by hungry children," said Frank. "They will look at your feet. If there's blood on your feet, they know you got a moose.

"We are trying the best we can to be true American citizens," he continued. "Last Saturday, we put on a dance for the March of Dimes and our whole village went to it and we collected $52.26."

Everyone knows that this is our land, he said desperately. And then he recited a story which was to be told over and over again in Alaska in the years to come. In 1937, Frank said, the people of Minto had been asked by the Bureau of Indian Affairs if they wanted to go on a reservation. The people knew very little about reservations but what they did know convinced them that they did not want one. The village rejected it. But, in connection with that reservation proposal, Frank's father, then the chief, had made a map of the Minto People's land, showing where they hunted and fished.

In 1951, the people filed a land claim with the Bureau of Land Management in Fairbanks. At that time the man who was then chief simply took a map and indicated with a circle what land the people used. Ten years later, in the wake of oil exploration around nearby Nenana, the Minto People revised their claim. This time Frank, who was chief, took his father's map to the BLM office and, using it, indicated what lands his people claimed. And now the state was planning to build a road across that land and had not even consulted the village about it.

"Now I don't want to sound like I really hate you people, no," said Frank. "If we were convinced that everyone would benefit, that the people of Minto would benefit, we might go along.

"The attitude down there [in Minto] is that you people were going to put a road into Minto Lakes without even consulting the people who live there, who hunt and fish there, who use the area for a livelihood."

He concluded, "Why do you propose to build the road only halfway, why not build it all the way to Minto? That way the village would get some benefit, everybody would be happy."

Everyone connected with the project protested innocence. Roscoe Bell, head of the Division of Lands, said that he knew nothing whatsoever about the Indians' claims. Governor William A. Egan hastily assured the Natives that the state had no intention of taking their land. Federal officials were more honest. The manager of the Fairbanks BLM office remarked that the Natives were "damn lucky to get five acres, let alone 160," a reference to the 160-acre homesteads Natives could acquire under the virtually unused Alaska Native Allotment Act, which the Interior Department was then hoping to spruce up a bit to satisfy the Natives' growing hunger for land.

The truth of the matter was that the state director of the BLM had been ordered to dismiss all claims unless they were to lands actually occupied by Natives. After a flood of Native claims, Minto's among them, were filed in 1961, the Interior Department's Regional Solicitor ruled that these claims involved so-called Indian title and that any settlement of them must involve a careful determination of facts. This was not the business of the BLM, he argued, so local land offices were instructed to dismiss the claims on jurisdictional grounds unless they involved 160-acre allotments.

The Fairbanks sportsmen, many of whom had been hunting at Minto Lakes for years, grumbled that they had never seen any Natives around there. The reason, although they did not know it, was that the BIA had advised the Natives against coming too near to white hunters in order to avoid some undefined "trouble."

The BIA and the BLM, both agencies of the Interior Department, were operating at cross purposes in Alaska in the early 1960s. On the one hand, local representatives of the BIA urged Natives to claim all the land they could under every available means and to protest state selections of land which they considered their own. On the other, local BLM offices dismissed these claims as fast as they were filed. Whether either agency had the authority to do these things was a point of controversy.

The state Division of Lands tried to arrive at an accommodation with the Minto People. Bell suggested that the state and the Indians could resolve their differences and work out a mutually satisfactory land use plan for the area. He proposed that the state select and patent the land and then sell parcels of it back to the Indians.

"No," replied Frank. "As long as I'm chief, we won't give up our land. We have the same idea the state has. The state wants to develop this land and that's our aim, too." And the village hired a lawyer, a young attorney who had just moved back to Alaska from Washington, where he had spent a number of years in the Solicitor's office at the Interior Department. His name was Ted Stevens. Stevens took the case for nothing.

Frank's fighting words were spread to every Native community in the northern part of the state by the lively five-month-old Native weekly, the *Tundra Times*, published in Fairbanks by an Eskimo painter named Howard Rock. Apropos of what was happening in tiny Minto, Rock wrote: "We Natives should realize that we will not be able to compete fully with big business for a long time yet. Since we cannot do that now, we should try to hold on to our lands because that is the greatest insurance we can have . . . without land, we can become the poorest people in the world." [31]

In March 1963, when that editorial was written, Rock's people were already among the poorest in the world. The median income of the village Natives was then $1,204; one in three had no cash income at all.[32] While most supplemented this pittance by hunting, fishing, and picking berries, the high cost of living in Alaska ate away what little money they had. In some villages, eggs cost 98 cents a dozen; stove oil, 41 cents a gallon; fresh fruit, when available, more than $1 a pound. The further north a man lived, the more he paid. Most Natives owned little in the white man's sense: perhaps their dog teams, hunting implements, and a umiak (an open boat used in the hunting of whales), but not the land on which their meagre houses stood.

Rock's people were among the most unhealthy too. Twice as many Native babies as white babies died in their first year. The average age at death was thirty-five. Ten times as many Natives as whites died of influenza and pneumonia, three times as many died of accidents; and twice as many killed themselves.

The Natives were also poorly educated. Most adults had had no more than an eighth-grade education. The young people had usually been to school longer but their education ill equipped them for life in either the village or the white man's world. Native children learned to read from Dick and Jane readers and, after elementary school, most had to attend boarding schools outside Alaska.

What the villagers had, or thought they had, were acres and acres of land, much of it land for which few white men had had any use in the past. What the Natives were discovering painfully in the early 1960s was that they did not have the most essential part of this land—the white man's deed.

There are few parallels in white western society to the Indian concept of land ownership. For Indians, there is no private land. It is held communally for all people to use. Men flow over it and enrich it for their descendants. The man is a part of the land and the land is a part of the man. As Rock wrote in March, "land that has been kind to one's ancestors and the same land that is kind to their descendants today is hard to leave, let alone give up." [33]

As the state embarked on its land selection program, it found itself in constant conflict with Native groups. The fear of losing their land aroused the Native people and eventually united and radicalized them. By 1968, even inhabitants of the most remote villages knew what was at stake. What happened in the villages in

Alaska in the 1960s was a revolution. It was the result of a number of happy coincidences, as most successful revolutions are, including the emergence of several young Native leaders who were capable of thinking simultaneously as white men and as Natives.

Much of the credit for politicizing the Natives must go to Rock and his *Tundra Times.* Rock helped his people bridge a period of economic and cultural crisis with pride and resilience and turned them into a potent twentieth century political force. In 1962, Rock was urging his people to choose carefully among available white men in the upcoming election. By 1970, his front page carried the announcement of the Alaska Native Political Education Committee's endorsement of one white and two Native Democrats for statewide offices.

The *Tundra Times*, which was born on October 1, 1962, grew out of a discussion at a meeting of northern Eskimos in Barrow the previous year. In November 1961, some Eskimos organized the Inupiat Paitot, "The People Speak," in reaction to an AEC plan to blast a deep water harbor in Arctic Alaska. One problem was how to bring Native news to the villages.

The Natives had no money for a newspaper. However, about this time, a retired Harvard Medical School professor, Dr. Henry S. Forbes, had become chairman of the Association on American Indian Affairs' Alaska Policy Committee and had visited the state. One of the people he met in Alaska was Rock. Forbes and Rock talked about the need for a Native newspaper and the possibility that a foundation might finance it. The Inupiat Paitot tried unsuccessfully to get a grant with which to start such a paper. Finally, Forbes agreed to back it modestly on the condition that Rock, who had had no newspaper experience whatsoever, would run it. Rock took the job.

However, he had no choice but to get experienced assistants, so he turned to a white reporter who had covered the first Inupiat Paitot for the *Fairbanks News-Miner,* Thomas Snapp. Snapp had expressed some interest in a Native newspaper at that time. When Rock first approached Snapp, the journalist was packing for a long-awaited trip Outside. He was interested in what Rock had to offer but he wanted to go on his vacation. However, Rock wore down Snapp's resistance and, two days before he was scheduled to leave, the reporter canceled his vacation and went to work for the *Tundra Times.*

In the first issue, Rock laid out his paper's aims. First, it was to serve as a medium in which Native organizations would be able to air their views. There were, at that time, several Native organizations in addition to the Alaska Native Brotherhood. Secondly, the *Times* would "keep informed on matters of interest to all Natives of Alaska." And third, it would be a forum for the "unbiased presentation of issues." It was to be nonpartisan, although it might endorse political candidates who showed a particular sensitivity to the Natives' needs or who were supported by Native organizations.

"Long before today there had been a great need for a newspaper for the northern Natives of Alaska," wrote Rock in his first editorial. "Since civilization has swept into their lives in tidelike earnestness, it has left the Eskimo, Indian and Aleut in a bewildering state of indecision and insecurity between the seeming need for assimilation and, especially in the Eskimo areas, the desire to retain some of their cultural and traditional way of life." [34]

Those first issues of the *Times* are an intriguing hodgepodge. Rock wrote a column about Eskimo culture. A professor at the University of Alaska contributed a column in Eskimo—with translation. Some of Rock's village correspondents recounted legends of their villages. Others wrote about successful hunts or natural catastrophes. One told of how the resident missionary had outwitted the game wardens when they came to the village searching for illegally taken moose. Another told how a whole village voted a month early in the general election because it was necessary for everyone to leave the village before the election to hunt caribou. And Rock editorially, although not always grammatically, urged the people to be proud, to stand up and take charge of their own affairs. In one editorial he wondered why so many young Native women were leaving the villages in search of non-Native husbands. In another he urged village leaders to attend a "sawmill clinic" the state was sponsoring to learn how to get a sawmill for their villages and then despaired when no village leaders showed up ("Does one have to give them a hot foot to get them to do something?"). He worried about the tendency of state and federal officials to bypass the village chief or council in favor of the BIA teacher or the local missionary when they needed information about a village. (This editorial elicited a letter from a reader who advised chiefs: "If Senator Gruening or some other VIP . . . comes to your village, put on your best clothes and meet the plane." If you're still bypassed, the letter-writer continued, have your council pass an ordinance that you and it will meet every official visitor.)

Letters to the editor or through the editor to white officials or other Native groups were most important in the early days of the *Tundra Times.* More than one controversy was hashed out in its Letters to the Editor column.

For instance, the Natives were able to force Senator Ernest Gruening, D-Alaska, to alter somewhat his position on how the Native claims should be settled. On April 15, 1966, the *Times* ran a story based on a press release from the Senator's Washington office in which he urged the federal government to pay the Natives cash to settle the claims quickly because, he said, the claims were based on the "dubious grounds of aboriginal rights." The BIA had encouraged these Native protests against state land selections, the Senator said, adding, "This situation is intolerable to the state of Alaska and constitutes a repudiation by fiat of an executive agency of provisions of the Statehood Act enacted by the Congress. In effect, the Department of the Interior has arrogated to itself the legislative function of Congress by its refusal to act on land selections filed by the state."

The story continued: "Senator Gruening recommended that the Department of the Interior should issue an order immediately refusing the acceptance of any additional Native protests to state land selections. He recommended further that the Secretary of the Interior should dismiss the protests now pending." Finally, the

Senator said that if the Interior Department considered the claims valid, it should introduce legislation to settle them immediately.[35]

The following week the *Times* carried a lengthy letter to Gruening from a student at the University of Alaska, Willie Hensley, an Eskimo from Kotzebue who was to emerge as one of the most articulate of the young Native leaders. Hensley wrote:

> Realizing that Congress has the ultimate power over what is done with Native lands, I feel that you should have at least made an effort to discover what the Natives desire behind these land claims before creating a prejudiced attitude toward them by your recent statements.
>
> . . . Now that we have developed and been civilized to the point where we can take legal action, we hope that we will be given the opportunity to press our claims and not have you deprive us of this right by prohibiting our "protectors"—the Interior Department—from accepting our protests against state land selections. If the claims are halted, what will be the basis for compensation for the large majority of Eskimos who have made no claims but are among the most needy in terms of material comforts and simple luxuries?
>
> . . . Compensation in cash would certainly be a simple and quick solution for Congress to buy off the Native claims, but it seems that we should be given the opportunity to voice our opinions on the matter.[36]

A week later, Senator Gruening responded in an interview with the *Times.* He said that he was agreeable either to legislation giving the Court of Claims jurisdiction over the Native claims or to separate pieces of legislation settling each of the specific claims and that he would introduce whichever type of legislation the Interior Department wished.[37] Only a few months later, Senator Gruening met with other members of the Congressional delegation and representatives of the Interior Department about the Native claims. However, there were no Natives present.

Throughout the years between 1962 and 1966, the Natives felt white Alaska pressing in upon them inexorably. The story of Minto was repeated over and over again. Alarmed by reports of the huge

Rampart Dam proposed for the Yukon River, which would flood much of northeastern Alaska, villages in the area began to file claims to millions of acres. Among them were the Birch Creek People, who, in September 1963, sent a map to the Interior Department along with an apologetic note: "Map is kind of small. We couldn't mark out just here and there because people move about looking for better hunting or fishing areas, things like that. And that's where their old grave or old village sites are. We could mark out a few of them." [38] All across northeastern Alaska, villagers met, discussed the dam, and decided they wanted no part of it. This annoyed the dam's chief booster in Congress, Senator Gruening, who was busy rebutting conservationist criticism of the project and did not want any opposition in his own backyard. Following one of a number of regional meetings on classifying millions of acres as a power site, the first step in the project, Senator Gruening angrily got into a plane and flew away from Fort Yukon, leaving representatives of the newly formed Gwitchya Gwitchin Ginkhye ("The Yukon Flats People Speak") standing impotently on the ground. The villagers filed more claims, one of which included land in Canada as well as in Alaska.

What happened in Tanacross, not far from Fairbanks, became a minor national scandal. Like the Minto People, the Tanacross villagers had sent a claim to the BLM in the early 1950s. They never heard from the BLM about it but wisely kept copies of the papers they had sent. In November 1961, almost ten years after the villagers had sent it off, the Tanacross claim inexplicably showed up in the Fairbanks BLM office, where it was rejected. The following year, the BIA filed a protest with the Secretary of the Interior on behalf of the village. Nothing happened, and by 1964 the state had selected land around Tanacross for itself. Thereupon Chief Andrew Issac filed a blanket claim to all the land around the village. This was where matters stood until the following year, when the Indians discovered that the state planned to sell lots around their fishing ground, George Lake, at Alaska's booth at the New York World's Fair. The villagers were outraged. Roscoe Bell, still Director of the Division of Lands, said he didn't know anyone claimed the land around the lake. However, an employee of the Division of Lands claimed to have been fired because he had called Bell's attention to the Indian claims and, when asked to keep quiet about it, had refused.

By the middle of 1968, almost 337 million of Alaska's 375 million acres were formally claimed by one Native group or another. The largest single claim was that of the Arctic Slope Native Association to 57 million acres, consisting of the entire North Slope and several potential oil fields.

As the size of the Native claims grew, the Natives also grew politically. They had the intoxicating experience of seeing themselves become a force in the state. The Natives had always had their own representatives in the legislature. In fact, it was disproportionate representation until the Supreme Court's 1964 reapportionment decision whittled the Ice Bloc, as the Native legislators were called, down to size. However, from 1966 on, the Natives began to acquire a new sort of political power that was potentially far more disturbing to white Alaska.

In part, this was the direct result of the federal War on Poverty. In the spring and summer of 1966, federal money from the Office of Economic Opportunity (OEO) was used to establish a nebulous program called Operation Grassroots. At its head was Charlie Edwardsen, whose ideas were considerably more militant than those of the established Native leaders. Edwardsen's job was to stir up community interest in the villages and he saw it as mandate to encourage the development of regional Native associations. He provided new struggling groups with seed money, the first time the villagers had gotten federal funds directly instead of through the auspices of the BIA. And, as these associations came into being, they blanketed the state with land claims.

The oldest Native association was the Alaska Native Brotherhood in southeastern Alaska. The ANB was founded in 1912, shortly after millions of acres of Indian lands were taken to form the Tongass National Forest. Over the years the ANB became quite sophisticated in land matters. By the early 1960s, the Tlingit-Haida claims case had been pending before the Court of Claims for nearly thirty years. One of the founders of the ANB had said to a young Tlingit, William Paul, in 1925, "William, the land is yours. Why don't you fight for it?" It was a revolutionary idea which young Paul mulled in his mind before suggesting it to the Brotherhood in 1929. Six years later, Congress passed legislation permitting the Tlingits and the Haidas to press their claim. Paul, by then a lawyer in Seattle, represented the Tlingits and the Haidas and later the Arctic Slope Native Association until he was in his seventies. His son, Fred, continued to do so throughout the claims battle in Congress. Between them they logged more hours of legal service than most of the Natives' other lawyers put together. Paul, an orator of the first order, was also something of a spiritual father to a generation of Eskimo militants.

Gradually other Native groups were organized in the 1960s, until in June 1963, the *Tundra Times* reported a proposal that three of them, the Inupiat Paitot, Dena Nena Henash ("The Land Speaks") and the ANB join together. Rock was enthusiastic. His report elicited an angry letter from Eben Hopson, a Barrow Eskimo and member of the legislature. Hopson wrote scornfully: "I can just picture you and a handful of other Eskimos sitting at a conference table with a battery of members of the Alaska Native Brotherhood and being voted down on every proposal you might have." [39] Rock hastily replied that what he had had in mind was not a "merger" but an "affiliation."

The Native groups were traditionally suspicious of one another. Each had different needs and different constituents. While the ANB had decided to go to the Court of Claims and get money for its lost land, the northern Eskimos did not want money. They wanted the land itself.

Since the early 1960s, the *Tundra Times* had urged the Natives to realize their political potential as one-sixth of the state's population and use it wisely. "[Natives] must make up their minds to participate in politics and form political units in their communities," Rock wrote in 1964.[40] Later, Edwardsen and his fellow Operations Grassroots workers toured the state, talking about a far more radical kind of political activity.

Traditionally, the Natives were Democrats. They voted as a bloc. Every Alaskan politician running for statewide office set aside a few days for shaking hands and eating muktuk (raw whale meat) in the villages. But no one discussed substantive issues with the Native electorate.

The first Native organization to go political was the Arctic Slope Native Association (ASNA), based in Barrow, the largest of the Native villages. In May 1966, before a primary election, the ASNA announced its support for four candidates for statewide offices, all of them Democrats: Governor Egan and Senator Bartlett, who were seeking reelection; Mike Gravel, a state legislator who hoped to unseat the incumbent congressman, Ralph J. Rivers; and Hopson, who was running for Secretary of State. But a few weeks later, the ASNA announced that its leaders would work actively against Senator Gruening, also a Democrat, when he faced reelection in 1968 as a result of his attitude toward the Native land claims. In a letter to the *Tundra Times*, an officer of the ASNA, Hugh Nicholls, wrote: "True, Senator Gruening has done some good things for the Alaska Natives [as Territorial Governor, Gruening had pushed through stringent antidiscrimination laws], but he should not consider them to be like faithful dogs who having been given a few pats and kind words remains [*sic*] wagging his tail, while his master swings a few well aimed kicks at his head. No, Senator, the Native people have begun to grow up, even as you have grown old [Gruening was in his seventies by then], and what they would accept gratefully without question 40 years ago, they will not do so now."[41]

Later that same month, the new Northwest Alaska Natives Association, which had its headquarters in Kotzebue, endorsed the same four candidates plus a local one, Willie Hensley, the Eskimo who had written to Senator Gruening. He had helped found the NANA and was running for a seat in the legislature.

This political activity stirred a response outside the Native community. The candidates who had not received Eskimo endorsements complained that they had been overlooked. All campaigned extensively in the villages. One of the pace-setters in this respect was Gravel, an energetic real estate developer from Anchorage who decided to go all out to get the Native vote. When Gravel almost toppled Representative Rivers, who had been the Congressman since statehood, in the 1966 primary, the Barrow Eskimos were jubilant. Representative Rivers was anathema to the Natives because he was reported to have remarked, apropos of Minto's plight, "What would they do with it [the land] if they had it? They wouldn't use it. It would just lie there." The ASNA endorsed Representative Rivers' Republican challenger in the general election, Howard Pollock. Pollock was elected in November.

In October 1966, representatives of all the Native organizations and many individual villages met in Anchorage to form the organization which was to become the Alaska Federation of Natives (AFN). Its purpose was laid out in the preamble of its constitution, adopted the following year:

> We, the Native people of Alaska, in order to secure to ourselves and our dependents the rights and benefits to which we are entitled under the laws of the United States, and the state of Alaska; to enlighten the public toward a better understanding of the Native people; to preserve the Native cultural values; to seek an equitable adjustment of Native affairs and Native claims; to seek, to secure and to preserve our rights under existing laws of the United States; to promote the common welfare of the Natives of Alaska and to foster the continued loyalty and allegiance of the Natives of Alaska to the flag of the United States and the state of Alaska, do establish this organization. . . .

Delegates to the 1966 meeting were not in agreement about everything, but about one thing they were unanimous. "It has now become necessary for the Native people of Alaska to make a determined stand to protect what is rightfully ours," read the policy statement of the Statewide Native Conference on Land. And it concluded, "The Statewide Native Conference on Land desires to have the land problems settled and further desires to have the development of Alaska continue." [42]

The Native conference in Anchorage was a political event. Candidates for state offices flocked into the hotel where the Natives were meeting and treated them to a succession of breakfasts, luncheons, cocktail parties, and dinners. The Natives, most of whom had little money, appreciated both the food and the irony. "One thing about this conference is that you don't have to spend money for meals," remarked Emil Notti, an Indian who later became the first President of the AFN.

Six months later, in the spring of 1967, a small group of Natives who had attended the Anchorage meeting the previous fall got together and formally set up the AFN. This led to an outcry from Eskimos and Interior Indians who were not present. Many had not been invited to the second conference, which met the same weekend as the Tanana Chiefs' Council, to which most Interior Indians belonged. Those not present at the second Anchorage meeting naturally suspected that the Anchorage Natives planned to take over the new organization. In addition to adopting a constitution, the second conference drew up a slate of candidates and it seemed to some Eskimos and Indians that there were too many Anchorage people running for statewide offices. Nevertheless, in the interest of unity, the northern Natives decided to go along with the AFN for the time being.

In the meantime, there had been an election in Alaska. The state's handling of its land selection program and the Native claims were now in a different pair of hands, those of Governor Walter J. Hickel. A brash hotel owner from Anchorage who, according to legend, parlayed 34 cents into several million dollars after arriving in Alaska, Hickel had done little during the campaign to indicate any particular sensitivity to the Natives. The theme of his campaign had been getting Alaska moving economically. Despite the oil discoveries in Cook Inlet and on the Kenai Peninsula and the establishment of pulp mills in the southeast, Alaska's economy was not booming. Hickel promised to bring industry into the state.

Before Hickel, the Egan administration and the Congressional delegation, all of them Democrats, had taken the approach that the land claims were a federal problem and that any settlement of them had to be initiated by the Interior Department. Late in 1966, Interior Secretary Stewart L. Udall had put a moratorium on the process of patenting state land selections in order to preserve the status of Alaskan lands until the Native claims were settled. This was the beginning of the so-called land freeze. His action followed Native protests against plans to sell oil and gas leases on the North Slope, on land which had been selected by the state and tentatively approved for patent to it. Hickel, who by this time had taken over as Governor, was determined that the sale should take place as planned. After the BLM halted it, the Governor ordered the sale to take place anyway, commenting, "Alaska is on its way to becoming one of the major oil-producing states of the Union and artificial barriers to development must be broken down for the benefit of all." [43] The sale was held.

Meanwhile, Native leaders tried to enlist Governor Hickel's support for legislation which would allow the villages to take their individual cases to the Court of Claims. But the Governor, realizing that any litigation would take years, opposed the legislation. Characteristically, he did not do it gracefully. Invited to attend a meeting with Native leaders to discuss the matter, the Governor instead sent a representative to tell the Natives that he opposed the bill. The Native leaders were insulted and furious.

However, the Hickel approach to a problem is a bit like tossing a dozen baseballs into the air and swinging at all of them on the theory that one swing will connect. Next, the Governor tried the courts. In February 1967, the state filed a suit to test the validity of the Udall land freeze. The test case involved land at the northern edge of Mount McKinley National Park, where the Nenana Indians lived. Most of the land around Nenana had been taken by the state under its land selection program but almost none of it had been patented. The Indians had filed a claim to almost three million acres. Eventually they were allowed to intervene in the state's suit.

Governor Hickel did not let it go at that. Later in February, he met personally with the same Native leaders he had earlier insulted and proposed the bare bones of a legislative settlement which called for granting the Natives full title to some land around the villages and surface rights to considerably more.

In May, after years of hedging on the claims, the Interior Department finally drew up a settlement bill for Congress. The essence of it was the grant of 50,000 acres to each Native village and the distribution of a small amount of money to all Natives. The Department would continue to control both the land and the money after it was given to the Natives. Somewhat reluctantly, because he did not like the control the Department planned to retain over the situation, Senator Gruening introduced the bill in June.

However, the settlement bill generated ill will even before it was introduced. In early June, Representative Pollock set up a briefing in Washington for Governor Hickel and other members of his administration at which Interior Department officials were to explain their proposal. Although Representative Pollock had promised that Natives would attend any such meeting, none were invited. State legislator Hensley and Emil Notti, who was by then president of the new Native federation, read about the forthcoming briefing in an Anchorage newspaper, borrowed money from the Cook Inlet Native Association, flew to Washington, and presented themselves at the Interior Department. Shortly after this briefing, the AFN met in Anchorage and voted to oppose the Interior Department bill and seek a Court of Claims settlement instead. At their request, Senator Gruening also introduced legislation which would make this possible.

In October the AFN met again, this time to hear Governor Hickel's Attorney General, Edgar Paul Boyko, urge cooperation between the Natives and the state against an old enemy, the

Interior Department. The state wanted to avoid lengthy litigation in the Court of Claims. The Governor was afraid that, unless the claims were settled quickly, the oil companies then doing exploratory work in the Arctic would leave. Three major companies had extensive leases on the North Slope—Atlantic-Richfield, Humble Oil, and British Petroleum—but only one was still actually drilling. Its rig was at work at Prudhoe Bay. Boyko told the Native leaders that the Governor was ready to take the claims seriously. He proposed compensating the Natives for land already taken from them on the basis of its value at statehood and giving the Natives full title to some other land as well. A few weeks later, the Governor named thirty-seven people to a Land Claims Task Force which was to write a mutually acceptable land claims bill. Representatives of the Interior Department were also to work with the task force.

In November 1967, Secretary Udall made a trip to Alaska and toured the state, meeting with non-Native and Native leaders. Several times during this trip, the Secretary suggested that revenues from offshore oil and gas leasing might be used to compensate the Natives. He said the idea had come to him as he flew from Washington to Anchorage. The proposal to use these revenues was like a twig thrown to drowning men. The Natives grabbed at it because Alaska has 60 percent of the nation's coastline and geologists suspect that oil and gas underlie much of it. (One reason Secretary Udall was in Alaska was to announce plans to offer leases in the Gulf of Alaska for sale.) The state grabbed at it because it did not mean taking a slice out of their own projected revenues for the immediate future. It is unclear why Secretary Udall suggested it. He must have realized that the Bureau of the Budget would never approve such a plan. But he also thought a large land settlement was a political impossibility. "I'm not making a proposal. I'm tossing out an idea," he cautioned.[44]

Meanwhile, the Senate Interior Committee announced that it would hold public hearings on the land claims in Anchorage early in 1968 and that it wanted to hear from the village people. Urged on by the AFN, the *Tundra Times*, and their attorneys, the villages began selecting spokesmen to send to Anchorage, drawing up maps of the lands they claimed, preparing other exhibits, and writing out their testimony. The Alaska Natives had come a long way. From a disorganized group of vaguely discontented but helpless people, separated from one another by almost unimaginable distances, formidable physical barriers, and ethnic differences, they had become a cohesive force, one to be reckoned with in the Alaskan political arena. The people were putting pressure on the government and the government was beginning to respond, or so it seemed at Christmastime 1967. The end looked deceptively near.

A winter road consists of ice and snow packed down firm to make a smooth surface. Here, a truck supply train moves toward Anaktuvuk Pass before construction of the new permanent road between the Yukon River and Prudhoe Bay.

Oil from wells like this one near Prudhoe Bay will be carried through 789 miles of pipeline to the ice-free port of Valdez in southern Alaska.

Since fall of 1969, 48-inch pipe has been stacked in Valdez waiting for construction to begin on the Trans Alaska Pipeline. (*John M. Berry*)

This tanker is small by comparison with those which will eventually be used at Valdez.

Empty fuel oil barrels are piled up at Deadhorse, the commercial airport for the North Slope oilfield. The smoke is caused by burning petroleum from a new well British Petroleum is testing. (*John M. Berry*)

The pipeline will cross the rugged Brooks Range at Dietrich Pass, now a wilderness haven for big game and smaller animals. (*John M. Berry*)

This young moose is eating succulent water plants in a pond in the Kenai Moose Range. (*John M. Berry*)

Huge tankers will travel through these narrow straits from Prince William Sound into Valdez Arm, on which the southern terminal of the Trans Alaska Pipeline will be located. (*John M. Berry*)

This is a typical scene in Barrow, the largest Native village. The pipe supplies natural gas from Naval Petroleum Reserve #4, which surrounds the village. (*John M. Berry*)

Laundry dries outside on a balmy June day in Barrow as the leads in the ice pack on the Arctic Ocean behind gradually widen. (*John M. Berry*)

These children are playing on a crude seesaw over an open sewer in Barrow. The balloon tires enable the vehicle to travel over the tundra with minimum damage. (*John M. Berry*)

This umiak was left on the beach beside the Arctic Ocean, following the Whale Festival at Barrow. (*John M. Berry*)

A huge earth-moving machine climbs a slope above the frozen Yukon River as construction begins on a 360-mile road along the northern section of the pipeline.

4
NATIVE AGAINST WHITE

Dear Mr.—

You make me sick working for the lazy, dirty natives. Perhaps you want to line your pockets. Don't see Russia or China helping them out. They put them to work.

Do you expect us whites that raised our families to keep Alaska going and clean up after these natives? I think decent people will move out and leave you with the natives. See what happens then.

Mrs. Disgusted

(anonymous letter to an attorney for a Native group, winter 1969–70)

February 8, 1968 was an historic day for the village people. They came into Anchorage from Barrow and Bethel and Kotzebue, from Minto and Unalakleet and Nenana and Nunivak Island, from English Bay and Grayling and Cantwell and Akiakhak, from the Kenai Peninsula and the Copper River Valley, from the southeastern panhandle and the Arctic Slope. They brought with them skin boats and harpoons and knives, the implements of their trade as fishermen and hunters, the symbols of their centuries of residence on and subsistence from the land. The land was what they came for—the land to which they felt a mystic bond. They came to tell three United States Senators and a Congressman how they felt about their land.

By eight A.M., Sydney Lawrence Auditorium in Anchorage was packed for the first Congressional hearing on the Alaska Native land claims. Senator Lee Metcalf, D-Mon., presided over the hearing in the absence of Senator Jackson, who had been detained in Washington. Senator Metcalf is a big, shaggy man with a deceptively genial manner. At his right sat Alaska's elder statesman, Senator Gruening, a spry elf of nearly eighty but more nimble-minded than many men half his age. On Senator Metcalf's left was Senator Paul Fannin, R-Ariz., a ramrod-erect conservative of impeccable tailoring. The Congressman was Representative Pollock, who was sitting in on the meeting. The Interior Committee met for three days, an hour earlier each morning, to accommodate the people, Native and non-Native, who wished to testify. The record of those three days is the record of a land and a people.

"I went to school when I was six or seven in Karluk," said Ewen Moses Laumoff of Old Harbor. "In 1903 or 1904, the whole village of Karluk had flu. We got the flu from the white man. So the school was closed. I was in the second grade. About 40 people died from the flu, my dad and mom died too. The school closed for about 22 or 23 years and by that time I was married and too old to go back." [45]

"Around 1890, Wales, Diomede and Shishmaref were among the villages most abused by the whalers," said Frank Topsekok of Teller. "Each spring the whalers would pick up many young women from these villages to be used for their pleasure during the whaling season. Sometimes they were not returned until the following year, and then not all came home. This caused great anxiety and grief to parents and husbands. It was not until many years later that law and order was obtained through the Coast Guard. At this time many of the women finally were returned to their homes." [46]

Charlie Franz of Port Moller told of an odyssey. "During the first part of the twentieth century, there was quite a bit of interest in coal, fisheries, and fur in the area that is now Port Moller to Cold Bay and a trader whose name I would like to withhold started a couple of trading posts in the area. There was only one problem, there were not enough people in the area at that time to make the venture profitable.

"I cannot even guess what sort of power of persuasion this gentleman used but he convinced the U.S. government in 1908 that it would be a wise and humanitarian move to bring a number of the destitute Eskimo from the northwestern part of Alaska to this fair land of milk and honey, or caribou, furs, fish and seals, if you prefer.

"The officials must have been very convincing because in 1909 the revenue cutter Bear arrived at Nelson Lagoon, near Port Moller, with about 40 families of Eskimos from North Cape, Little Diomede Islands, Nome and other villages from the far north. This was not the limit of their persuasion. They painted such a bright picture of the Promised Land that a number of skin boats with whole families aboard sailed and paddled their way from the Arctic to the Alaska Peninsula. This journey took them two heartbreaking years. For a while they prospered for this land was rich with furs, game and fish, but not rich enough to support this huge influx of people.

"Then fate took a hand. Tragedy struck in 1918 and 1919. This was the influenza epidemic that followed World War I. Whole families and villages were wiped out completely and, when it was finally over, there remained only parts of four families. By 1930, these people had either died or moved away. There was only one Eskimo of all those heroic people who had put so much faith in our government. That Eskimo was me, for my mother was one of those who had made that historic trip." [47]

"Of the 15 children that I started school with at Akiakhak," recalled Harold Harvey Samuelson, "the rest died of tuberculosis." [48]

Said Jerome Trigg of Nome, "Ooyolik, my mother, may be one

of the only Natives that ever got any compensation from gold mining in Alaska. She had a sod hut at the mouth of Daniel's Creek at Bluff, where over a million dollars in gold was taken out. After the miners moved Ooyolik's igloo and her two sons, they gave her a poke of gold. This was undoubtably a very small compensation for all the gold that was taken out of the area she occupied. Tony Tony of Nome Village received 25 pounds of flour, sugar and tobacco for showing white men Anvil Creek, which led to the discovery of the Nome gold fields." [49]

"We have, according to the government, absolutely no rights on Nunivak Island because it was declared a national wildlife refuge in the 1930s," said George King, who lived there. "The island has apparently been set aside for ducks, musk ox and reindeer. We have not even been able to get a townsite and, according to the Executive Order establishing the reservation, we are not even there. There are approximately 10,000 reindeer and 600 musk ox on the island. These animals have eaten up most of the vegetation and deprived our people. We are only allowed to take five reindeer per year per family. It is hard for us to understand why the government reserved all of Nunivak Island for the animals and left none of it for the people." [50]

Through an interpreter, Joe Seton of Cooper Bay told the legislators, "He came from Cooper Bay, one of the oldest villages on the coast, on the western coast. That's where he came from. Very old, actually don't know the village when it started. Started long time ago like he says back when the earth was thin, very thin. Now, you people asked me to come here to testify and prove that it is my place where I was born, where my grandparents were born and lived there. He emphasized, 'It's my place.' " [51]

And Ernest Kignak of Barrow said, through his interpreter, Eben Hopson, "The fact that he is now in his later years—getting kind of old to be hunting like he used to, but he does have grandchildren and children and he feels that if the land is given to the Eskimos in a small package, you might say, he can see the problem of not being able to go out further to follow the migration of animals up there that is available for hunting. And he has this feeling of not having enough land for his children and grandchildren to hunt in." [52]

"A man without a title under his feet, he loses something," said Claude Demientieff of Galena, "He doesn't have authority." [53]

There was also a hint of the opposition which was to grow in Alaska in the coming years. Initially, it came from the mining interests. George Moerlein, representing the Alaska Miners Association, spoke at the hearing. "Gentlemen," he said, "I submit to you that neither the U.S., the state of Alaska, nor any of us here gathered as individuals owes the Natives one acre of ground or one cent of the taxpayers' money." [54] He based his argument upon a peculiar interpretation of the Treaty of Cession. He said that the additional $200,000 the United States paid Russia in effect guaranteed their title to Alaska against all future claims.

Nor was there any moral obligation to the Natives, according to Moerlein. There were some Alaskans who doubted the Natives' legal footing but argued that the United States had a moral obligation to settle claims equitably, though cheaply, since they were not legally valid. But Moerlein argued that the BIA and the Public Health Service had spent enough on the Natives' welfare "to the specific exclusion of the Negro, Oriental and Caucasian residents of Alaska" to more than compensate for lands taken from them.[55]

Moerlein tried to prove that a settlement involving land grants and any kind of mineral royalty, one of the ideas being considered by Governor Hickel's land claims task force at that time, would destroy the infant state. His argument was that land granted to the Natives would be closed to development, like national parks. In addition to restrictive regulations on oil, gas, and hard rock mineral exploration, a royalty would make the already high cost of exploration in Alaska even higher. And the best hunting and fishing areas would be closed to use by white men because they would belong to Natives. Moerlein said that such a settlement would strangle the state and that the Natives, too, would suffer.

Moerlein was not alone. A geologist named Lum Lovely agreed with him. Lovely said he based his opinions on a privately published book about the land claims by the former lobbyist for the Seattle "fish trust," W. C. Arnold of Ketchikan. But Lovely made a mistake by quoting Arnold to one of the architects of statehood, Senator Gruening. At one point, Lovely read from Arnold's book, "You can only conclude that this bill stands to be the biggest threat to Alaska's economic wellbeing as has come along since statehood." Alert at the first mention of his old archenemy, Senator Gruening interrupted, "Do you consider statehood a threat?" "No, no," replied Lovely, adding that he had gotten only four hours of sleep the night before and his "context" was bad.[56]

But the opposition to the Natives was not unanimous. Phil Holdsworth, a former Commissioner of Natural Resources who had recently given up the state job to become a lobbyist in Juneau for the Alaska Miners Association, took the opportunity to publicly challenge Moerlein's authority to speak for the miners as a whole. "I don't believe the group will agree with the position that there is no moral or legal right in this matter," said Holdsworth, who had not intended to testify. "Certainly the legality—legal responsibility is always subject to court action but the moral responsibility is certainly there." [57] He was fired a few days later.

The task force appointed by Governor Hickel the year before unveiled its plan for settling the claims at the hearing. This plan included a land grant, a money grant, and a royalty on future mineral revenues, the first time such a proposal had been offered. This meant that the state would be participating in the settlement (since most mineral revenues went to the state government), something not previously contemplated. Governor Hickel was asked if he was prepared to seek legislation providing for the state's participation and he said he was.

Specifically, the task force proposed that the Natives receive 40 million acres of land, although it avoided dealing with the touchy questions of which 40 million acres and whether the Natives would get it before the state had finished picking its 102 million acres or afterward. However, Governor Hickel did not intend for the state to be cheated out of any of its land and he hinted that perhaps the best thing would be to let the Natives take federally withdrawn land which the state could not take anyway, thus preserving Alaska's patrimony.

The task force proposal also gave the Natives a ten percent royalty on mineral revenues from the Outer Continental Shelf, the idea Secretary Udall had tossed off the previous fall. In addition, the task force proposed a grant of $20 million with the caveat that the federal government would be repaid this amount eventually with other Outer Continental Shelf revenues. And their proposal called for an unspecified financial participation by the state in the form of a royalty on mineral revenues. The Hickel administration agreed to this, to the delight of the Natives, who had viewed Governor Hickel with hostility before his election and during the early days of his administration.

And so the hearings ended on an upbeat note. The state government was cooperative. The Senate Interior Committee was returning to Washington to see what it could do for the Native people. Senator Jackson, its Chairman, was no longer just a name but a flesh and blood man who had walked down from his podium to examine the harpoons an old man from Bethel had brought with him to the hearing.

But 1968 was a year of tremendous upheaval in the United States. The country was involved in a pointless, costly war in Southeast Asia and was rent with domestic strife. The President, Lyndon Johnson, was defeated before he even ran for reelection by a war he could not win but would not lose. Civil rights leader Martin Luther King was murdered in Tennessee and riots erupted in more than a dozen cities after his death. Presidential candidate Robert F. Kennedy was killed in a Los Angeles hotel following his victory in the California primary. And Chicago policemen beat young antiwar demonstrators in the streets while the Democrats were choosing their candidate for President. The Alaska Natives were forgotten. Even had 1968 been an ordinary year, there was probably little chance that Congress could have resolved the claims quickly. Too many difficult questions had to be answered. To compound the problem, oil was discovered that spring at Prudhoe Bay and suddenly Alaskan land had a new value, both as potential oil land and as vanishing wilderness.

When the land claims did come under serious Congressional scrutiny in 1969, the cast of characters had changed considerably. There was a Republican administration in Washington. Governor Hickel had become the Secretary of the Interior. Although he was almost a legend in Alaska, Senator Gruening was defeated by Mike Gravel, who owed much of his victory to the votes from the Native villages. Another Alaskan legend, Senator Bartlett, gold miner, newspaperman, and territorial delegate, had died in December, and Governor Hickel had appointed Ted Stevens to fill Bartlett's seat for the next two years. Hickel himself was replaced by a friendly man well known in Juneau for singing "Your Cheatin' Heart" in the bar of the Baranov Hotel after long dismal sessions of the legislature, a practice Keith H. Miller continued as Governor.

The situation had changed, too. A United States District Court in Alaska had declared Secretary Udall's land freeze illegal, a decision the federal government promptly appealed. Undaunted by the court's action, the Secretary had expanded the freeze just before he left office, withdrawing for two years all unreserved public lands in Alaska in order to protect the Natives' rights. This superfreeze prevented any transactions on federal lands, from the patenting of state lands to the awarding of homesteaders' titles.

In December 1968, Governor Hickel, not yet confirmed as Secretary of the Interior, was asked about the superfreeze. He retorted, "Anything Udall can do by Executive Order, I can undo." The AFN was disturbed by this remark and by the state's crash program to complete its mineral selections by the end of the year. (Congress had adjourned in October without extending the deadline for making mineral selections another five years; this omission was largely a result of the Alaska Congressional delegation's ineptitude.) Therefore, the AFN withheld its endorsement of Governor Hickel for his new job. Since Hickel was encountering increasing opposition from all over the country, he needed help wherever he could get it.

In early January, while Governor Hickel was quietly visiting Senators to get their support for his nomination and a delegation of Alaskan businessmen was making the rounds of Capitol Hill offices to counter conservationist opposition to Governor Hickel, the AFN sent its own lobbyists to Washington. The five-man delegation, four Native leaders and their white lawyer, had two objectives. First, they sought a promise from the future Secretary that he would continue the land freeze. Second, they wanted members of the Senate Interior Committee to hear their side of the land claims story.

The Natives were successful at getting Governor Hickel to promise to continue the freeze. On January 17, 1969, the second day of his long, contentious confirmation hearings, the Governor promised to retain the freeze for two more years or until the claims were settled, whichever occurred first. He also promised to consult the Interior Committees of Congress before issuing any modifications of the freeze to accommodate public needs, that is, for roads, airports, and—most importantly in view of the Arctic oil discoveries the previous year—pipelines.

The Governor's promise came after more than three weeks of intensive pressure from the Natives. Once Govenor Hickel agreed to keep the freeze, AFN President Notti issued a press release endorsing him for the Secretary's job. Because of the Natives' insistence on the importance of the freeze, the claims became an issue during the Hickel confirmation hearings. Several members of the Senate Interior Committee, and most importantly Senator Jackson, made a point of asking the Governor about the land freeze, and the Committee, like the AFN, struck a bargain with the nominee: his retention of the freeze became one of the conditions of his confirmation. The hearings focused national attention on the problems of the Alaska Natives for the first time.

Shortly after Governor Hickel became Secretary Hickel, the Federal Field Committee for Development Planning in Alaska completed a 565-page study of the Native land problem, the first of its sort, and made some suggestions about how to handle it. Senator Jackson, who had commissioned the report, and the late Senator Bartlett, who was more or less the father of the Field Committee, had wanted a document to point to when anyone complained that the subject needed further study. But they did not want anyone to read it too closely; hence its format, that of an elongated telephone book, and its bulk. Most importantly, the Senators, both of whom had helped to draft the 1958 statehood legislation, did not want the administration or the Interior Department to do the study, fearing that any such study would recommend gutting the Statehood Act to settle the land claims. Both wanted the state to get its 102 million acres of land. Having the Field Committee, created in the aftermath of the 1964 earthquake to help plan redevelopment in the state, do the study was really an end run around the administration.

As might be expected, the Field Committee under Joseph H. Fitzgerald kept the Statehood Act firmly in mind as it considered what to do about the Natives' claims. But the committee was also concerned about integrating the Natives into Alaskan society and giving them a stake in the state's economic development. The report warned against the creation of racially defined enclaves and recommended that land be given to individuals or legally recognized communities, not tribes. The committee recommended the following: a land grant of between four and seven million acres plus the right to use other public land for subsistence purposes; $100 million for lands taken in the past; and a ten percent royalty on mineral revenues from public lands in Alaska for ten years. Having received the committee's report, Senator Jackson set a date in late April for new hearings on the claims.

The AFN prepared for the hearings by hiring new lawyers. Former Supreme Court Justice Arthur Goldberg, then a member of the New York law firm of Paul, Weiss, Goldberg, Rifkind,

Wharton, and Garrison, became interested in the Natives' case through his son, who lived in Anchorage. When the AFN first approached Goldberg to see if he would take their case, the former Justice was worried about whether the AFN was really a unified group. His apprehension was well founded, as it turned out, although he concluded at the time that the Natives were unified. Later, former Attorney General Ramsey Clark also joined the firm and took an active role in the case.

Goldberg first sought the assistance of a Washington lawyer familiar with both the Interior Department and "Indian" law. He selected Edward Weinberg, who had served as Solicitor of the Interior Department from April 1968 to February 1969 and had been Deputy Solicitor for five years before that. After leaving the Department when Hickel became the Secretary, Weinberg had joined the law firm of Wyman, Bautzer, Finell, Rothman, and Kuchel, in which former Senator Thomas Kuchel of California, once the ranking minority member of the Senate Interior Committee, was a partner.

Goldberg's choice of Weinberg created an uproar in Alaska, where local lawyers, who had had the Natives' business all to themselves, charged that it was illegal for the former Solicitor to take such a case. The question was whether Weinberg could legally represent private clients on a matter in which he had participated while serving in the Interior Department. As Deputy Solicitor, Weinberg had taken part in a conference between former Secretary Udall and Edgar Paul Boyko, then Attorney General of Alaska, at which the state had protested the first land freeze. Later, as Solicitor, Weinberg helped to draft the Department's 1967 land claims bill. And he had advised the Secretary on the public land order withdrawing all unreserved public lands in the state, the 1968 superfreeze. Eventually Assistant Attorney General William H. Rehnquist ruled that Weinberg could participate in the case and Weinberg agreed to drop out if litigation ever arose.

In Alaska, envious local lawyers spread rumors about the size of the fee Goldberg would get for the case. The former Justice took great pains to stress that he was taking the case as a public service, but this did not stop the rumors. He began to get letters and telegrams from various lawyers with Native clients. These attorneys claimed that they were being treated shabbily by Goldberg and his associates and that they were being kept out of "secret" meetings between Goldberg and the AFN. The telegram Goldberg received from John W. Hendrickson, who had been representing the village of Unalakleet, was typical. "I am surprised to learn of the manner in which your office is dealing with the Alaska counsel," he wired. "Your office appears to be interfering with my contract and those of other attorneys. Also, you appear to be soliciting our clients." [58] Goldberg informed the AFN that he would resign. "I am in receipt of several letters and telegrams from certain lawyers purporting to represent affiliates of the AFN," he said. "These communications are entirely lacking the respect owing one who has served our country in three of its highest offices, and I shall not dignify them with a direct reply." [59]

The AFN board hastily convened in Anchorage to set its house in order. Goldberg had wired Notti that he would reconsider his resignation. The board voted on whether or not to accept the former Justice's resignation but the vote ended in a tie. Then the board voted again and agreed to reconsider it. Several weeks later, after arrangements were made to write a schedule of attorneys' fees into the settlement, Goldberg said he would go back to work for the AFN.

As a result of the whole affair, the Natives developed an instinctive distrust of lawyers, both Alaskan and Outside. They wondered, with good reason sometimes, about their local lawyers' devotion to the Native cause since many of them also represented clients with mineral leases or held leases themselves. Later the Natives were to wonder how much good their well-known Outside legal talent did them.

The Senate hearing on the land claims began before the AFN's internal troubles had been resolved; therefore, the Natives and their lawyers had no positive recommendation to make when they testified. It was not until late June 1969 that the Federation drew up its claims bill. In the meantime, the Interior Department had proposed a settlement which differed considerably from the task force plan Hickel had endorsed as Governor. The Interior Department proposed to give the Natives about 12½ million instead of 40 million acres of land and $500 million instead of $20 million, but no royalty or other sort of revenue sharing.

At a meeting with the Secretary and members of the Alaskan Congressional delegation, the Natives distributed copies of a June 20, 1969 AFN memorandum outlining what was to become the Federation's position on how the claims should be settled. The memorandum proposed that the Natives should get the following: full title to 40 million acres, to be allocated among the villages according to their size; $500 million; a two percent royalty in perpetuity on all revenues from all other public and state lands in Alaska; and the creation of a statewide Native development corporation and up to twelve regional development corporations to manage the land and money received in the settlement.

It took a while for the true dimensions of this proposal to sink in in Alaska. On the surface, it seemed not unlike the recommendations of Governor Hickel's task force. Actually, it was quite different. When this difference was finally understood, a howl of indignation arose from special interest groups that could be heard from Juneau to Fairbanks.

First, there was the land. The AFN proposed to give each village fee simple title to four townships, or 92,160 acres. This meant mineral rights as well as surface rights. But most private citizens in Alaska could not acquire mineral rights anywhere. Also, the AFN proposed that the Natives be allowed to pick their land anywhere in the state, even if the land had already been selected by the state and tentatively approved for patent to it, like the Prudhoe Bay oil field.

Then there was the two percent royalty. Since under the statehood provisions Alaska receives 90 percent of the revenues from mineral leases on federal lands and the federal government gets the rest, the royalty would come either from Alaska's share, thus reducing it to 88 percent, or from the federal government's, thus reducing it to eight percent. Two percent of all mineral revenues from state lands would also go to the Natives. And the royalty was to be open-ended, thus giving the Natives "a perpetual interest in lands which had been theirs for thousands of years."[60]

And there were the development corporations. Each Native would be given stock in them. The idea was that the charters of these corporations were to be very flexible so that they could make the most of the resources they had—the land, the mineral rights to some of the land, and the revenues provided by the settlement. The villages would be able to convey their lands to their local regional corporation or the state corporation, provided the regional corporation agreed, and the AFN anticipated that most villages would do so. The mineral rights on all village land would be conveyed to the regional corporations automatically and the revenues from any minerals found there would be divided 50–50 between the regional corporation controlling the rights and the other regional corporations. The $500 million in federal money and royalty revenues were to be distributed on the following basis: 75 percent to the villages (where only 20 percent would be given out on a per capita basis and the rest would be used for public projects); 20 percent to the regional corporations; and 5 percent to the statewide corporation.

The AFN proposal was innovative, designed to circumvent the problems which had arisen with Indian claims settlements elsewhere. It avoided a per capita dsitribution of cash, which could be squandered by some individuals, and which, in any case, would be too little to help most Natives. It also retained the Native concept of communally held land but adapted it to changing times. While the present generation of villagers, and even their children, might continue to live off the animals and fruits of the land and the fish of the oceans, a day was coming when subsistence living would no longer be possible, even in the villages. Then the land would have to sustain its people in a completely different way.

But it was also a proposal which trod on many sacred

toes—the federal government's, the military's, the conservationists', the special interests', and the state's.

Some 85 million acres of Alaska were special federal reserves. In the case of some, like the Naval Petroleum Reserve at Point Barrow, powerful committees of Congress objected strenuously to giving up any portion of the reserve to private individuals. In the case of others, like the Arctic National Wildlife Range in northeastern Alaska, conservation groups resisted any effort to dismember it. The AFN proposal would have allowed the Natives to acquire land in either Pet Four or the Arctic Wildlife Range if they wanted it.

There was no unanimity within the federal government over the use of reserves to settle the claims. The Interior Department recommended using all federal reservations, including Pet Four. (Alaskans and the oil industry had long been trying to open up Pet Four.) But the Forest Service, which had extensive holdings in the southeast, objected to using the forests in the settlement.

The conservationists were anxious that the Natives should get as little federal refuge land as possible and that some guarantee be provided that none of the land would pass out of Native hands. At this stage, conservationists hoped that land would be given in trust to the villages so that the Secretary of the Interior would continue to manage it. But, if there was one thing the Natives agreed on, it was that they wanted control over their own destinies, not a trusteeship by the Secretary of the Interior, however benevolent. While the Sierra Club patronizingly called the Natives "good conservationists" and "the aboriginal stewards who kept the land in what we regard as virgin condition," it was plainly unwilling to take a chance on their stewardship in the future.

The oil and gas industry had its own concerns. In 1969, the smaller companies were worried because the Federal Field Committee had suggested that competitive leasing ought to be required throughout the State, not just in areas known to contain oil. Senator Jackson agreed with the Field Committee on this. The independents, or nonintegrated companies, were upset because this could mean bonus leasing, which would require large outlays of capital at the time the leases were sold. The independents feared they would be unable to compete with the major oil companies if bonuses were required. Ten Canadian independents—the Asamera Oil Corporation Ltd., Banff Oil Ltd., Home Oil Company, Husky Oil Ltd., National Nickel Company Ltd., Ranger Oil Ltd., Scurry-Rainbow Oil Ltd., Sunlite Oil Company, Ulster Petroleum Ltd., and Western Decalta Petroleum Ltd.—had a special problem. All had filed for noncompetitive federal oil and gas leases in various parts of Alaska between March 1967 and November 1968. The lease applications, which covered roughly 20 million acres, had cost them collectively about $3 million. Because of Native protests and the subsequent land freeze, the BLM had not processed these applications routinely, as it would otherwise have done. The Canadian companies feared that Congress would approve a claims settlement which required competitive leasing and that their applications would then be rejected. They wanted language in the settlement which protected their priority rights. So they hired a Washington lawyer, Max Barash, once counsel to the United States Geological Survey, to lobby for them.

Another oil industry group which was interested in the outcome of the claims settlement was the Western Oil and Gas Association, represented by John H. Pickering of Wilmer, Cutler, and Pickering. The association is a trade organization whose members are active in six far western states, including Alaska. They wanted to make sure their members' existing rights, that is, federal and state mineral leases, were not changed. While most of the proposed settlements provided that land grants to the Natives should be "subject to existing rights," the oil companies were afraid that this would not be enough protection. They wanted their rights spelled out in the settlement legislation.

Like the state itself, the oil companies were anxious that tentative approval for patent, as well as actual patenting, be considered an "existing right." The state had already leased some 2.4 million acres, most of them on the North Slope, which were only tentatively approved for patent to it by the BLM at the time the freeze halted all further consideration of state land selections. And the state planned to offer more leases on tentatively approved land at the September 10, 1969 lease sale. The settlements proposed by the Interior Department and the Field Committee recognized tentative approval as a valid right, but the Natives argued that their claims took precedence over the Statehood Act and that they should be allowed to take tentatively approved land if they wanted it. This raised the question of how valid state oil and gas leases on these lands were, a question of great concern to even the major oil companies. The discovery of a huge oil field at Prudhoe Bay provided an additional incentive for settling the claims, but in the summer of 1969 the oil and gas industry was not too worried about what would happen.

The state government was worried. First of all, the state wanted to make sure that all the land it had selected to date—about 26 million acres—would remain in its hands. The state argued that tentative approval constituted a jurisdictional transfer since the state was then able to lease the land (as it had already done on the North Slope). The state's second concern was that the mineral rights to any land given the Natives remain under state control. The state's position, as enunciated by Governor Miller in August, was that the Statehood Act had given the state control over all leasable minerals (primarily oil and gas) and that to give these rights to any individual or corporation would be a violation of that compact.

Then there was revenue sharing. The various settlement proposals differed as to whose revenue was to be shared. In April, before the AFN had suggested a two percent perpetual royalty, Governor Miller had said guardedly that his administration endorsed the "percentage sharing principle." In August, with the AFN proposal before him, the Governor reversed himself. He said, "I cannot . . . support a revenue-sharing proposal such as the two percent overriding resource royalty that is advocated by the Alaska Federation of Natives. The $500 million compensation figure is commensurate to the Native losses and I can see no justification for additional compensation. . . . Furthermore, the two percent proposal conflicts with the Statehood Act and the province of the Alaska State Legislature and I have no authority to support such a proposal." [61]

The Governor's remarks indicated a return to the old position that the claims were a federal matter entirely and that funds to settle them should not come out of Alaska's pocket. In 1967 and 1968, the state under Governor Hickel had agreed to contribute. In fact, in 1968 the legislature voted $50 million toward a settlement, although the funds were contingent upon lifting the land freeze within six months. In the Natives' eyes, the state, under its new Governor, was going back on its word.

Governor Miller's stand reflected the thinking of certain powerful groups in the state and their influence upon him and other members of his administration—business groups, Chambers of Commerce, trade organizations, sportsmen's groups. The attitude of the Alaska Sportsman's Council was typical. Its members opposed any settlement involving large grants of land to Natives because "the first time that a white citizen goes out for his annual hunt or fishing trip in his traditional hunting and fishing areas and is told that he is trespassing on Indian land will signal the dawn of a new era in Alaska." [62]

In the early summer of 1969, when the AFN first presented its settlement proposal, the opposition was disorganized. The Governor's testimony before the House Interior Committee in August helped to pull it together. Later, Governor Miller took an even more intransigent stand and the various special interest groups rallied around him. Their theme was that the AFN proposal was an Outside scheme to deprive Alaska of its resources foisted on ignorant Natives by scheming Outside lawyers. The leaders of the opposition played skillfully on not-so-latent racism, bigotry, and greed to orchestrate it to a fine pitch. Much of the responsibility for the frenzy of hatred generated in Alaska during the closing months of 1969 must be placed upon the Governor and his advisers for their inflexibility and lack of sensitivity, upon organizations like the State Chamber of Commerce for their ill-timed attempts to influence Congress, and upon the *Anchorage Daily Times* for its editorials, which fueled the fire. Their actions radicalized many Natives and produced schisms within Alaskan society which may take generations to heal.

On October 18, just before the House Indian Affairs Subcommittee was scheduled to hold public hearings on the claims in Anchorage, the *Daily Times* published an editorial which said some of the provisions of the AFN proposal were "shocking," particularly the perpetual royalty. And "that is only one of a basketful of fantastic proposals advanced to the Congress by the native association and its lawyers, headed by Arthur Goldberg of New York," continued the *Times.* The editorial was entitled "The Goldberg Bill."

"Under the terms of Mr. Goldberg's proposals," continued the *Times*, "approval of the native land claims settlement bill would erase—so far as Alaska only is concerned—the federal mineral leasing laws; the state mineral leasing laws applied to land selected by the state under the Statehood Act but not patented because of the Udall Land Freeze; the federal mining laws; the homestead laws; and all other federal laws respecting the disposition of public lands in Alaska." [63]

The bill would have the effect of "sealing off" 97 percent of Alaska, said the *Times*, because the Natives would be able to take land from anywhere, including areas previously selected by the state—Prudhoe Bay and the North Slope lands leased the month before for $900 million—and because this land would all be withdrawn from entry under the public land laws until the Natives had finished making their land selections.

The newspaper continued:

> There are many more stunning provisions in these 85 pages, but the total effect would be to cripple the development of Alaska for all its citizens.

> Alaskans, of all people, seek justice for the natives and seek a society undivided by racial animosity and united in a common goal of building a great state for all the people—natives, whites, Alaskans, those who were born here and those who came here to find a new way of life.
>
> Alaskans want the aboriginal claims of the native people satisfied to the extent that is just and proper, to the extent that the government must settle an obligation with justice to all the people of the U.S. This debt is not one of the state of Alaska to a segment of its citizens. What debt there is is a federal responsibility.
>
> And that is what the Congressional committees must realize, and that is why this Goldberg Line that has been fed into the Alaska Native Federation bill is a threat to all the people of the 49th State.[64]

"Gutter journalism," snapped a visiting Congressman at the public hearing. Representative Ed Edmondson, D-Okla., said the editorial contained gross exaggerations and misrepresentations. And he was furious at the *Times*' insinuations that the proposal was a Jewish scheme foisted on dumb Indians. The *Times* took up the gauntlet, smugly.

"Mr. Goldberg indeed is the chief counsel for the native group and the man to whom the natives look to prepare and plead their case," the paper replied on October 21. "This seems not to be gutter journalism but merely the reporting of a fact." [65] Racism had nothing whatsoever to do with the newspaper's stand, the second editorial continued. It was entirely a question of whether or not the state should pay for the settlement.

"Most Alaskans feel, we believe, that the settlement of any native aboriginal claim is a responsibility of the U.S. government, against which the claim is filed.

"Representative Edmondson apparently believes that Alaska—alone of all the states, including Oklahoma—must participate in the financial settlement.

"But where is the justice in that?" [66]

And a third editorial the following day contained this argument:

> The claims filed by the native people were not filed against the state of Alaska. They are filed, and rightly so, against the government of the U.S.
>
> It was not the state of Alaska which appropriated what the Indians and Eskimos contend is aboriginal land. If seizure is a legal fact, it resulted from an act of the federal government.
>
> State selection of land has been made under provisions of a law passed by the Congress of the U.S., in the form of the Alaska Statehood Act. That act represents a compact with the people of Alaska, one that must be honored.
>
> The unity of Alaskans, therefore, must be directed toward Washington and a just and proper settlement on behalf of the federal government for our native citizens.
>
> What debt and obligation there is lies with the federal government and the Congress has a responsibility to resolve it as the representative of all the states.
>
> When that is made clear to Congress, by the natives and by all Alaskans, we can unify behind legislation which will settle the lands [sic] claims with justice to all.[67]

These editorials reflect the thinking of Robert B. Atwood, longtime publisher and editor of the *Times* and the man who protested in 1950, during Congressional hearings on Alaskan statehood, that he could see no connection between statehood and the Native land claims.[68] Atwood considers himself an architect of statehood. He was chairman of the Alaska Statehood Committee, a public relations effort to counteract the lobbying activities of Outside interests during the statehood fight. Atwood devoted his life to statehood for almost a decade. Alaska is his baby.

Atwood speaks for the colonists, the men and women who came to Alaska when it was just a poor territory, made it a state, watched it grow, and prayed that nothing would halt that growth. On the tenth anniversary of statehood, Atwood could write: "Statehood has done—and still is doing—everything its advocates said it would. The success of the first ten years shows no sign of diminishing in the next ten years. . . . Economically, the dynamics of growth and prosperity have never been more widespread and prevalent throughout Alaska."[69] He could write this without mentioning the state's close brush with bankruptcy and the onesidedness of its economy and particularly without mentioning the oppressive poverty of the Native villages.

It was the golden prospect of prosperity which motivated the Alaska Statehood Committee. But in the years that followed statehood, men like Atwood saw Alaska stymied at every turn. Frustrated and unable to comprehend what was wrong, they turned to a favorite Alaskan myth which the state's remoteness encourages—the Outside conspiracy. In the past, an indifferent federal government, overzealous national conservation groups, and exploitative Outside interests had helped to explain the pitfalls on the way to Alaskan prosperity. Now, on the eve of oil prosperity, the Natives and their "New York lawyers" were the culprits.

Atwood supplemented his editorial stand by publishing sporadic articles by W. C. Arnold, the self-proclaimed expert on the land claims. Arnold's articles often appeared in the *Times* with headlines like "Goldberg Bill Would Tie Up the Best Land." The Arnold articles were a startling mishmash of misinformation and innuendo. It is small wonder that people who mistook them for facts were frightened and angry. It is a sad commentary on the state of the Alaskan press that few learned otherwise. The *Times*' competitor, the *Anchorage Daily News*, tried to straighten out this information gap by publishing in the fall of 1969 a lucid explanation of the various settlements that had been proposed as well as a list of suggested reading for Alaskans which broadmindedly included the works of Arnold. But too few Alaskans read the *Daily News*.

The *Times* articulated the views of a powerful clique in Alaska represented by the Chambers of Commerce throughout the state. Atwood gave these views respectability by putting them into print. He was, one Native leader remarked bitterly, appealing to the "dark side of human nature." His editorials opened the floodgates, and many non-Natives, from bankers to homesteaders, began to say publicly what they may previously have thought or said only privately.

Alaskans wrote their Congressman and Senators angry letters protesting the "giveaway." The number of letters was disturbing and they came at a crucial point. The Senate Interior Committee was about to begin mark-up sessions on the settlement.

In 1969, a mark-up session was a private committee meeting (many are now open to the public) where members and staff drafted the legislation to be presented on the floor as the committee's bill. In some cases, it is a routine meeting at which minor modifications and technical amendments are made to a piece of legislation on which most members already agree. In other cases, this is the point at which the bill is actually drafted. In the fall of 1969, there were three proposals for settling the claims before the Senate Interior Committee but none were acceptable to all members of the committee or to all parties to the dispute. What the Committee had to do was to take parts from each, modify them, fit them together, and come up with a bill which, at least, would not lead to law suits. The spectre of thirty years or more of litigation hung over the proceedings. (It had taken more than thirty years to settle the Tlingit-Haida claims.) Senator Ted Stevens remarked that he doubted if he would live to see the end of such litigation once it began. He was then forty-six.

On November 18, just as the Committee started its mark-up session, it received a letter from Governor Miller. The Chambers of Commerce clique had won its first victory with the Governor. Miller, a pleasant but not particularly strong-minded man, was faced with a dilemma. He had never been elected Governor in his own right and, if he wanted to be, he faced a tough campaign in 1970. The clique was his natural constituency, so he did what they wanted. He sent a letter to Senator Jackson which dispelled any doubts about where the state stood.

After a preface stressing the need for a "fresh approach" to the claims and alluding to the strong opposition to the AFN proposal in the state, the Governor plunged to the heart of the matter. Four townships for each village was "completely unacceptable and unrealistic," he said. The villages ought to get no more than two townships (46,080 acres) apiece at the very most and many should get less. Furthermore, the amount of land withdrawn for the Natives to make their choice from should also be no more than two townships apiece. In other words, the Native villages were to be given no choice of land at all. Congress was to decide what they would get.

Governor Miller wanted no state land used in the settlement. Regardless of the status of its patent, if the state had picked the land, that land was to be the state's. The Natives could not have it. And no mineral rights were to go with any of the land given the Natives.

There must be no two percent royalty. "We are unalterably opposed to any form of revenue sharing, overriding royalty or other provisions affecting revenues presently being received by the State of Alaska under provisions of the Alaska Statehood Act," wrote the Governor.[70] And he sent his Attorney General to Washington to argue that such a provision was illegal and unconstitutional.

Finally, Governor Miller wrote: "I must stress that a great majority of Alaskans are concerned over the unrealistic bill put forth by the Alaska Federation of Natives and their attorneys. I can only consider it as unacceptable, and regret that false hopes have been raised among our native peoples by the eastern attorneys representing them." [71]

The State Chamber of Commerce promptly began to organize support for the Governor's stand. The day after the Governor sent off his letter, twenty business leaders met in Anchorage to discuss organizing an "Alaskan lobby" in Washington. They were told that the AFN already had representatives in Washington, which was true, and that the state had paid their way, which was not, strictly speaking. (Some of the money came from local antipoverty organizations, which of course had some state money.) The Association on American Indian Affairs had loaned the AFN

$35,000 so that President Notti and his family could be there and John Borbridge, Jr. was "on loan" from the Tlingit-Haida Central Council. The business leaders also threatened Senator Stevens, who would have to stand for reelection the following year, with dire political consequences. The Senator told them not to meddle.

In Washington, the Committee was having troubles of its own. Its two Alaskan members took up most of its time debating their different views of what a proper settlement should be. Senator Gravel favored the royalty. Senator Stevens did not. But the Stevens-Gravel debate was a luxury neither they nor Alaska could afford. And it raised issues which were better not raised, from the state's point of view, such as Alaska's 90 percent share of all federal mineral revenues. One Senator threatened to offer an amendment on the Senate floor giving his state 90 percent of all federal mineral revenues, too, and two percent to all resident Indians if the royalty provision was included in the settlement. Another suggested dropping the legislation right then and there and letting the Natives go to court.

Senators Stevens and Gravel had arrived in Alaska at about the same time to pursue political careers. Senator Stevens had been Solicitor of the Interior Department briefly during the Eisenhower administration. Senator Gravel had made and lost a good deal of money as a real estate developer on the Kenai Peninsula and around Anchorage. Both lost the first time they tried for statewide public office. After that first defeat, so Alaskan political legend has it, the two men had a conversation about their ambitions and how they intended to pursue them. Gravel is supposed to have said that he intended to find a rich man to pay his political bills and head straight for the top. Stevens is supposed to have said he thought he would start at the bottom and work his way up. In fact, both eventually served in the legislature. By one of those ironic twists of fate, both men chose to run for the Senate in the same year, 1968. Gravel defeated aging Senator Gruening in the Democratic primary with the help of two millionaire backers. Stevens narrowly lost the Republican primary to a wealthy Anchorage banker, who was in turn defeated by Gravel in November. Then Senator Bartlett died suddenly and Governor Hickel appointed Stevens to succeed him. So Gravel and Stevens found themselves freshmen Senators together. As might be expected of two such different personalities, the men disliked each other intensely.

However, Governor Miller's uncompromising letter drove them into an alliance. Senator Gravel wanted to catch the next plane to Alaska and have it out with the Governor personally. Senator Stevens remarked sadly that "mass hysteria" seemed to have taken over in Alaska. So they put aside their differences to work out a compromise which the Natives, the state, and the Interior Department would be able to accept. In this venture they had the backing of Senator Jackson, who was fed up with the bickering within his Committee.

On Saturday, November 22, the two met in Senator Gravel's office for three hours and hammered out a tentative compromise—a two percent royalty on mineral revenues to go to the Natives for a limited time. Then Senator Stevens flew to Anchorage, where he addressed the local Chamber of Commerce on Monday. He scolded the members roundly for their unwillingness to listen to reason. In doing so, he took a chance with his own political future. Like Governor Miller, Senator Stevens' natural constituency was the Chambers of Commerce.

Arguing for a compromise on the royalty, Senator Stevens told his listeners, "If you want to assume the $50 million annual burden of education, welfare, and health services, then be intractable. Don't listen to any compromise suggestion that might lead us out of this problem. But if you feel as I do that the two percent concept is a difficult one to swallow but one we're going to have to find some way to accommodate then listen to what Senator Gravel and I are trying to work out." [72] His audience received his words coldly.

The *Times* summed up their attitude succinctly when it said the following day:

> Realistic or not, Alaskans are getting fed up with efforts by Congressmen from other states to run our affairs and are

> beginning to wonder why we seem to have no response in the Senate and House to this meddling, interference and harassment.
>
> Many Alaskans would rather fight these intrusions rather [sic] than roll over and play dead, or rather than jump through hoops to satisfy the whims of Congressmen and Senators from other states.
>
> Clearly Sen. Stevens and presumably Sen. Gravel, believe the state's best interest will be served by compromise. They want a native settlement bill passed as soon as possible as an essential step to the lifting of the land freeze and to the approval of a right-of-way permit for construction of the Trans Alaska Pipeline System.
>
> They no doubt reflect a prevailing mood in Congress.
>
> Gov. Miller, on the other hand, reflects what seems to be a rising tide of sentiment in Alaska—and that mood is not in accord with the viewpoint of our senators.
>
> The Governor wants the land claims issue settled by Congress as a federal obligation.
>
> He wants the freeze lifted because it illegally restricts Alaska's rights to land selection under the Statehood Act.
>
> And he wants senators and congressmen from other states to quit meddling in Alaska's sovereign affairs, and apparently thinks our Congressional delegation should be making that message clear to their colleagues.
>
> So do we.[73]

Senator Stevens went right on plugging. The next day, he addressed the Anchorage Rotary Club. "I am here because I feel and I have felt for the last ten days from the tone of my telephone calls and letters and telegrams that there is a feeling developing here at home that Mike and I are wrong, that we should be opposing the Natives' claims. That's serious to me. . . . because I think it is unhealthy to have what I consider the non-Native community turning against the Native community at the very time after 102 years that we have the real clout to get this thing done." [74] He remained in Alaska the rest of the week, meeting with civic groups and Native groups, talking about the claims and urging unity, compromise, and restraint. But he received no assurances from either Governor Miller or the AFN that they would accept the compromise he and Senator Gravel were drafting.

On his return, Senator Stevens and Senator Gravel met again to discuss a compromise, this time with Senator Jackson and the Committee's ranking Republican, Senator Gordon Allott of Colorado, an outspoken opponent of any generous settlement, royalty or no royalty. Chairman Jackson emerged from this meeting beaming with confidence. The Committee would be through with the bill by Christmas, he said, thanks to the compromise. Although it was still a secret, the senators' compromise would give the Natives: $500 million in Congressional appropriations; title, including mineral rights, to between five and ten million acres around the villages plus use of another 40 million acres for subsistence purposes; and a two percent royalty for a specified number of years.

A few days later, Senator Gravel briefed the Native leaders on the compromise at lunch. Notti was noncommittal but it was clear that the compromise did not satisfy the Natives.

On December 11, the Alaskan Senators were scheduled to brief the full Committee on their plan. The Committee met, listened to the proposal, and arranged to meet again early the next week. But time was running out in 1969. Congress was expected to adjourn for the Christmas holidays on the 19th. As it turned out, the Committee met once more that year and it was not a very satisfactory meeting.

Senator Jackson went to the last meeting of the year angry because Senator Gravel had told reporters after the Thursday meeting that the Committee was in agreement and would meet daily in order to finish working on the bill. Senator Jackson was also furious because details of the compromise had appeared in an Alaskan newspaper. The Senator felt it was his privilege and duty as Chairman to make announcements about the Committee's plans and reveal the substance of its discussions. In an unusual move, he asked the Committee to vote him the sole spokesman for the group, which it did. The prerogatives of a Committee Chairman are taken very seriously on Capitol Hill so the Senator's anger was not surprising. Within his Committee, a Chairman can be all-powerful, if he chooses to be, and generally remains so as long as he does not abuse that power. Senator Jackson is generally a fair Chairman. Also, Senators Gravel and Stevens were both freshmen Senators, new boys who were expected to toe the line in the Committee. Instead, they had taken up a lot of time with their own argument. This had irritated Senator Jackson. The leak to the press was the last straw. In his role as sole spokesman, Senator Jackson told reporters waiting outside the committee room that the Committee planned to disregard all the settlement proposals previously made and start from scratch, taking each element of the land claims separately, discussing it and writing a totally new bill. The Committee was back where it had been in the summer and the year was gone. The Senators and the lobbyists went home to Alaska for the holidays.

The anti-Native backlash had radicalized many Natives, particularly the people on the North Slope. The Arctic Slope Native Association there had filed the largest single claim in Alaska. Their land was of great importance to them and they were already beginning to feel pressure from the oil companies at work there. There were reports of an oil company's landing barges at the site of a Native home and cemetery and taking the property over for its own use, with the help of a permit from the State Division of lands.

The more radical Native leaders in turn pushed the AFN further and further to the left. This was a slow process. But in February 1970, under pressure from the more militant members of the AFN, President Notti made what was for him a very radical speech to the Small Tribes of Washington. He called for a Native Nation in northwestern Alaska.

"If Congress cannot pass a bill that we think is fair," Notti told the Indians, some of whom later loaned the AFN money with which to carry on its lobbying activities, "then I will recommend a course of action to our statewide board of directors that we petition Congress and the United States to set up a separate Indian nation in the western half of Alaska. That area is ninety percent Native anyway, and will not get any non-Native settlers until there is something discovered that can be exploited." [75]

The more militant Native leaders were jubilant. Said Eben Hopson of the Arctic Slope Native Association, "It took him [Notti] a long time to express himself in this manner, after a couple of years of advocating a settlement, and my own feeling is that they [the AFN leaders] took all this time to perhaps fall into the same line that the ASNA has been advocating all the time." [76] Hopson also predicted that if the Natives did not get what they wanted from Congress in 1970, the AFN would elect more militant leaders and take a new approach to the settlement. The mild-mannered Notti's newly acquired militancy did not do much good in helping him retain control of the AFN. In March, Hopson became its Executive Director, and while Notti remained President in name, he lost his power. Hopson became the leader of the AFN.

On the other side of the spectrum, Governor Miller did nothing to change his intransigent stand on the claims. He continued to insist that his position was both generous and fair.

The administration in Washington also rejected revenue sharing, chiefly because the Bureau of the Budget objected to anything so "uncertain." As Interior Secretary Hickel wrote Senator Jackson in February, "We are . . . opposed to . . . a two percent overriding royalty in addition to the $500 million." [77]

However, some members of the legislature wondered if the state should not do something to indicate its willingness to participate, if only as a gesture. On the ferry *Malaspina*, en route to Juneau for the opening of the 1970 legislature, members of the Legislative Council, a leadership group, discussed what should be done. Republican State Senator Lowell Thomas, Jr. urged the legislators to take some sort of action. "I think Congress will forget it if Alaska doesn't make an effort," said Thomas. "Not only would it be encouraging to Congress, but it would close the door to suits against the state." [78]

But others disagreed. "In a poker game, you don't play your hand ahead of time," said one.[79]

The discussion on the *Malaspina* went round and round. In the end, the consensus was that the state did have some responsibility and that the claims had to be settled. One representative said, "I am willing to spend more than they [the Natives] are entitled to just to end the thing forever." [80] However, the legislature backed off from appropriating state money as part of the settlement after somc members of the Legislative Council were advised that to do so would be illegal, since it would aid one racial group in the state at the expense of another.

In Washington, Senator Jackson's Committee began meeting in February of 1970 and promptly got into a hassle over whether the Natives should get $500 million in cash, previously the least controversial element of the settlement. Meanwhile, the House Interior Committee was waiting to see what would happen. Back in October, the House Indian Affairs Subcommittee, accompanied by the ranking members of the full Committee, had visited Alaska. The Congressmen toured the state, visiting villages and holding hearings in most of the major towns. Their tour, coming as it did right after the exciting $900 million oil lease sale on the North Slope, got relatively little publicity. Chairman Wayne Aspinall, D-Col., had warned that this was a working tour and that the Congressmen did not want to be bothered with entertainment. Their Alaskan hosts took him at his word. But the lack of attention paid to the subcommittee took its toll in the end.

Representative James A. Haley, D-Fla., was the chairman of the subcommittee. At seventy-one, Representative Haley was an angular, white-haired man who had been president of Ringling Brothers and Barnum & Bailey Circus at the time of the terrible Hartford fire. Haley had been chairman of the subcommittee for years and was proud of the Indian legislation his panel had produced, although he was not a strong figure on the Interior Committee and acted as something of a cipher for Chairman Aspinall. But Representative Haley was very sensitive about his position vis-à-vis the powerful members of the Interior Committee and terribly conscious of his own prestige.

Apparently, Representative Haley was angered by the lack of attention he and the subcommittee got when they were in Alaska. He was annoyed that Senator Edward M. Kennedy, D-Mass., and his nonlegislative Indian Education Subcommittee, which had toured Native villages the previous spring, had gotten more attention than the House group. He decided that Alaskans did not care whether the claims were settled or not.

No one realized this at the time. All Representative Haley would say publicly was that he was waiting to see what the Senate would do. But he was not asked about it very often. In their concern about the Senate, everyone—the Natives, the Chambers of Commerce, the state—had forgotten about the House subcommittee. No lobbyists came near Representative Haley until the Senate had finally passed a bill in midsummer 1970. By then it was too late. The old man refused to see any of them.

"The lobbyists will write this bill over my dead body," he growled.

5
OIL COMES TO ALASKA

"If Alaskans don't make something big out of this country, they can be sure, at least, they will make a big fizzle."

Anchorage Daily Times,
July 30, 1957

". . . the essential question is: what the hell do we want to do with the state of Alaska? An approach is what we're looking for—a base. We can be a model state or the laughing stock of America."

Democratic State Legislator
Gene Guess, July 1969

"Hell, this country's so goddam big that even if industry ran wild, we could never wreck it."

Henry Pratt, adviser to
Governor Keith H. Miller,
spring, 1970

Shortly after the Richfield Company found oil on the Kenai Peninsula in July 1957, Alaskans waited in line outside the Anchorage Land Office to file applications for leases in the area. In true frontier style, they were ready to gamble on anything and oil was no exception. People had been playing the Land Office game ever since the early fifties, when the major oil companies first showed an interest in the Territory, but not on the scale they did that July. Forty-eight hours after information leaked out about Richfield's find, 1,015 applications for leases were filed on about 100,000 acres. The Bureau of Land Management literally had more business than it could handle. "We were geared for about 500 or 600 [applications] a year," commented a BLM official later, "and suddenly we had 17,000 applications." [81]

Eleven years later, when Richfield, now merged into an international oil giant, the Atlantic-Richfield Company (ARCO), cautiously let out word of what it had found in the Alaskan Arctic, there was a similar rush on the Fairbanks Land Office. The town had been seething with rumors of what was going on up there since late 1967. ARCO's tentative description of its find was the best thing that had happened to Fairbanks since the Gold Rush.

But, even as word spread of ARCO's "highly significant" find in the Arctic—an oil field conservatively estimated in July 1968 to contain five to ten billion barrels of recoverable crude oil—the state legislature debated whether to halt all oil production in Cook Inlet, not far from the first Richfield well, until more stringent antipollution legislation had been approved. In the midst of this hassle, the state arrested the captain of the tanker *Rebecca*, chartered by none other than ARCO, for dumping oily ballast in the inlet. And everyone in the state was arguing about whether to raise the production tax on oil. As this concern for the consequences of oil development indicates, Alaska had come a long way since the summer of 1957.

Geologists had long suspected that there was oil in Alaska. The first white explorers detected oil seeps in the Arctic. In the 1920s, there was a small producing field and refinery at Katalla, near the mouth of the Copper River. After World War II, the United States Geological Survey, which had been slowly tackling the task of surveying the Territory, decided to concentrate upon locating formations which were favorable to petroleum. The Survey's work led to a report which pointed out the attractiveness of several sedimentary basins and resulted in a startling grid of ten-foot wide swaths across the Alaskan wilderness, cut to facilitate airborne surveys, which are still visible today. At the same time, the Navy was doing exploratory work in Pet Four. While they located small quantities of oil and gas, their work in the Arctic was discontinued in the mid-1950s without conclusive results.

As a result of the USGS survey, oil companies began limited but steady exploration in the more accessible parts of the Territory. The federal government encouraged the companies by offering them incentives—leases at half what they normally cost on federal land. Exploration in Alaska is costly, even in the relatively temperate parts of the state where the companies began their explorations. As a result, it was chiefly the major oil companies which sent men to Alaska: Shell, Phillips, Standard of California, Humble Oil, and Union Oil of California.

The number of unexplored, easily accessible oilfields in the United States has been dwindling since World War II, and with them the traditional role of the independents, or smaller nonintegrated oil companies. In the past, the major companies have let the "little men" take the risks of exploring new fields. But the high cost of drilling in Alaska (ARCO spent $4.5 million on Susie Unit #1 near the Sagavanirktok River on the Arctic Slope before abandoning it as dry) kept out most of the smaller companies in the beginning.

Therefore, it is ironic that the first Alaskan discovery was made by a relatively small company, Richfield. In 1957, Richfield struck oil at Swanson River on the Kenai Peninsula. It was a small well but it set off a big boom.

Alaskans of all descriptions stood in line to file their applications to lease. The cost was not great—just a $10 filing fee and the first year's rent, then 25 cents an acre. Or a person could top-file on land already leased in hopes that the earlier lease would be invalid and the individual or company holding it would have to buy the top-filer out for many times the filing fee in order to obtain a lease that was valid. For those with neither the money nor the inclination to go it alone, there were operations like the Koslosky Development Company, which advertised in the classified sections of the Anchorage newspapers: "OIL: A Cooperative Venture in Alaska." The Koslosky Development Company, which was later to give Interior Secretary Hickel a few shaky minutes during prolonged confirmation hearings, consisted of seventy-eight persons, each of whom had filed for a single lease on 2,560 acres, then deeded it back to the corporation, so that if oil were found on any of the leases, all would share equally. In 1969, at the Senate Interior Committee's request, Governor Hickel got rid of his one-seventy-eighth interest in a one and one-quarter percent royalty on 600,000 acres, all that remained of the Koslosky Development Company. In fact, the Kenai oil boom produced few millionaires and the Alaskan economy was soon in its chronic doldrums again.

However, the timely discovery of oil gave statehood the push it needed, although the oil companies outwardly maintained a hands-off attitude toward the whole business. This was just so much acting. The industry had been discussing Alaskan statehood for some time and was already at work to get a favorable tax structure in the state. In 1955, the territorial legislature imposed a one percent production tax on oil, even though there was no oil being produced. The tax remained that low until 1967.

In Alaska the oil industry encountered a rather different situation than in other oil states, a situation which made it doubly important for the industry to establish a good public image. Almost no land in Alaska was privately owned in 1957, and by 1959 no mineral rights could pass from the state into private hands. The state was the mineral landlord; therefore, there was no group of oil-owning landholders who could be expected to work to keep taxes low or restrict the output of productive wells to protect marginal ones. The oil belonged to all the people, so the industry had to convince all the people that what was good for oil was also good for Alaska.

Alaska became an oil state slowly. In fact, the new state lived on hope until 1960, when Richfield and a partner, Standard Oil of California, found oil at Soldatna to the south of Richfield's original well. Then came the discovery of gas and oil offshore in nearby Cook Inlet. Altogether, fifteen oil and gas fields were found in the general area during the decade after Richfield's Swanson River strike. Oil and gas production rose steadily in the early 1960s, leveling off during the middle of the 1960s and then soaring in 1968 and 1969 to ten or twelve times what it had been in 1961. This was entirely the result of activity in southeastern Alaska. The North Slope has yet to produce any oil at all. However, during the 1960s, the major oil companies began to investigate the Arctic.

The development of the industry was spaced out over about ten years and change was so gradual that major confrontations over delicate matters like raising the taxes did not come until the eve of the discoveries at Prudhoe Bay. By then, according to the *Oil and Gas Journal*, a trade publication, Alaska ranked seventh among the oil-producing states, although it had only ninety producing wells.

While a single wildcat was being drilled up in the Arctic during the winter of 1967–68, the legislature was grappling with proposals to raise the severance, or production, tax on oil. There had been a disastrous flood in the Interior in 1967, resulting in tremendous destruction and property loss around Fairbanks. Governor Hickel asked a special session of the legislature to double the one percent severance tax in order to set up a disaster fund to cope with situations like the Fairbanks flood in the future. The extra one percent would cease when the fund reached a certain size. Officially the industry took no position on the increase, but both the Governor and the industry believed that the boost would offset anticipated demands for further hikes in 1968. The special tax was approved.

However, when the legislature met in January 1968, several bills raising the severance tax as much as an additional two percent were introduced. Governor Hickel argued that Alaska's entire tax structure needed studying and that no single part of it should be tampered with until such a study had been made. Rather than quietly acquiescing to a tax hike as it had the year before, the oil industry sent an elite corps to Juneau to deal with the threat. Like most state legislatures, the Alaska legislature as a whole knew almost nothing about the oil business, despite the industry's growing importance to the state's economy. The usual relationship between oil lobbyists and state legislatures is a curious one. Legislators must rely upon industry spokesmen for most of their information and advice while at the same time being under pressure from them to make decisions favorable to the industry.

In February 1968, legislative committees began hearings on bills to raise the severance tax. Their star witnesses were representatives of the Alaska oil industry. One man, K. C. Vaughan, a senior vice-president of the Union Oil Company of California, a corporation that had been on the Alaskan scene for some time, presented the industry's case. Vaughan told the legislators that to raise the tax would discourage exploration. Exploration in Alaska was almost prohibitively expensive anyway, he said. The industry had poured more than $1 billion into oil exploration in the state since the mid-1950s, according to Vaughan. Then he listed some figures which no legislator had the expertise to question.

Shell didn't want drilling

Finally, industry representatives showed a documentary film made by the Shell Oil Company about the hazards and costs of finding and producing oil in Alaska. Shown one night during a Joint Finance Committee meeting, it began with the sounds of a shrill wind whipping through a forest although not a bough on the screen moved. The legislators were skeptical.

They were even more skeptical when Vaughan produced figures projecting the state's share of oil profits over the next few years at some $39 million more than the state government foresaw. Whereas the state was estimating oil revenues of about $5 million for that fiscal year, Vaughan projected $7.6 million. By fiscal 1972, the projections were even further apart. The state thought it would make slightly over $8 million from oil taxes and royalties; the industry envisioned itself paying $18.8 million that year. The discrepancies led to demands that the industry disclose its profits from its Alaskan operations. The industry refused. Despite admonitions not to "kill the goose that lays the golden eggs," [82] the legislature in April raised the severance tax to four percent, thus creating a combined levy of 16.5 percent on oil production. (Alaska required a 12.5 percent royalty on all oil and gas leases in addition to taxes.)

Throughout all this, everyone was thinking about the North Slope. In January 1968, Atlantic-Richfield and Humble, who had leased Prudhoe Bay jointly, cautiously let it be known that they had found a "substantial flow of gas" there. "It's not a small amount," remarked the local exploration manager in Anchorage, but added, "This is a very rank wildcat." A few days before the Joint Finance Committee began its hearings on severance taxes, the company announced that it had struck oil. There were no estimates of the size of the find, just the cryptic comment, "It looks extremely good."

These words held for Alaskans the illusive promise of prosperity, power, and glamor. And nowhere was this more true than in Fairbanks, a small stagnant Alaskan city still living on its collective memories of the Gold Rush. Fairbanks likes to call itself the "Golden Heart" of Alaska. In reality, it is an unprepossessing collection of bars, pawnshops, and secondhand stores specializing in Arctic gear. Many of the streets are unpaved.

When oil was discovered 300 miles to the north, a sort of mass hysteria swept through the city. It was not so much a matter of individual land speculation, although there was plenty of that. Instead it was a wave of irrational fear that the city and its 21,000 people would somehow be deprived of their slice of petroleum pie unless they did something to attract the oil industry to Fairbanks. The city fathers, the local legislators, the members of the Fairbanks Chamber of Commerce, and the local newspaper were all frantic to get Fairbanks into the fraternity of oil towns, but no one knew what the city ought to do to promote itself.

In early 1968, everyone in Fairbanks was speculating about

what was going on on the North Slope. In March, ARCO had announced that, following an "inconclusive" four and a half hour flow, its new Prudhoe Bay well had flowed at the rate of 1,152 barrels a day. Furthermore, ARCO was drilling a second well (this was Sag River #1, the confirmation well) and there were persistent rumors that a third well was in the making. "We're a long ways from tall," said an Anchorage oilman, but nobody believed him. Fairbanks started planning for the windfall. When ARCO announced, on July 18, just how huge the North Slope oil find was—9.6 billion barrels according to the reputable petroleum consulting firm of DeGolyer and MacNaughton—there was pandemonium in Alaska's "Golden Heart."

One common assumption was that a big pipeline would be built to carry the crude oil from Prudhoe Bay to an ice-free port in south central Alaska and that the pipeline would go through or near Fairbanks. The city fathers speculated about whether such a pipeline could be tapped to provide fuel oil for Fairbanks, where the winter is long, dark, and bitterly cold. That would mean a refinery of some sort. Many people also assumed that a smaller pipeline would be built from Prudhoe to Fairbanks to supply the city with natural gas (there was such a gas pipeline from the Kenai Peninsula to Anchorage). The city officials began discussing setting up a gas distribution system that would use liquefied natural gas then but Prudhoe Bay gas later, and went so far as to talk to a Canadian company about it. Their idea seems to have been that if they had a gas distribution system, the oil companies would deliver the natural gas as a matter of course. Local businessmen pushed for improvements in the Alaska Railroad, which came to a dead end at Fairbanks. At the same time, Governor Hickel started talking about extending the railroad all the way to the North Slope.

In August, ninety members of the Calgary (Canada) Chamber of Commerce visited Fairbanks at the invitation of their Alaskan counterparts. The trip had actually been planned before anyone knew the size of the oil strike on the North Slope, but the scenario read as if someone had known. Calgary itself had grown from a sleepy little village into a boom town following oil discoveries in Alberta Province in 1947. The message from the Calgary businessmen to their Alaskan hosts was this: start immediately to woo the oil companies to Fairbanks. Don't depend upon being the closest city to the new oil discovery. "Distance means nothing to the oil companies," advised one Canadian. Another urged going "all out" to help the industry. He advised Fairbanks to show them that the city had a plan for making their business and their lives easier.

Since the publisher of the *Fairbanks News-Miner* was an active member of the Chamber, the newspaper gave the Canadians' visit front-page treatment and the editor was assigned to write stories about the meetings that ensued. On August 10, 1968, Editor Murlin Spencer wrote:

> They said it also is of utmost importance that Fairbanks assure the oil companies it will have the kind of community where people like to live. Not only land and industrial sites are necessary, but Fairbanks must offer the amenities as well—golf course, good clubs where oilmen can meet, good hospitals and medical facilities, good schools, play areas.[83]

And he followed the story up with this editorial:

> Send out an emissary now—to talk with the leaders of the oil industry, to tell them we want them to establish their headquarters in our town; to assure them they will have our complete cooperation down to the most minute detail; to tell them of our city and our plans for it.[84]

This is exactly what Fairbanks did. The Chamber of Commerce formed an oil impact committee which at birth was galvanized into action by reports of a conspiracy among Anchorage businessmen to keep the oil industry out of Fairbanks. The *News-Miner*'s publisher was a member of the committee and naturally its activities got good play in the newspaper.

The Fairbanks Industrial Development Corporation was set up to lure the industry into Interior Alaska. Its chief booster was the mayor, H. A. "Red" Boucher, a former naval officer who rose to relative fame by winning a television quiz show. The mayor was bursting with ideas. He publicly castigated an airline for raising the cost of a round trip air freight charter from Fairbanks to the Slope from $2,700 to $3,100. He recommended that a delegation go to Calgary to see how that city had handled its oil boom. He urged merchants to seek out North Slope oil companies needing supplies since the companies were too busy to shop around for themselves. Then, in November, he took Borough Chairman John Carlson with him to Anchorage for a parley with ARCO and Humble Oil officials there. However, Boucher's real coup boggles the imagination. He and several other city officials were going to New Orleans in December to see if they could get Fairbanks named an "All-American City." The mayor decided that the delegation would return by way of the oil capitals of Houston, Dallas, and Los Angeles.

By the Alaskans' own account, the oilmen were startled but they responded gracefully, wining and dining the visiting Alaskans at local Petroleum Clubs along the way. Boucher was enthusiastic. "We feel we have moved right through the heart of the Petroleum Industry," he said upon his return.[85] The industry, given an opportunity to win the hearts and minds of the local citizenry that it does not often get, did a good job. The Alaskans returned convinced that Atlantic-Richfield and Humble Oil, the partners at Prudhoe Bay, would do everything they could to help the community. In fact, the oil companies had already been quietly at work on their public image there. Fairbanks was trying to raise money for a new hospital and some of the more sizable gifts to it came from oil companies. Later, an oil company helped to finance new buildings at the nearby University of Alaska. Everyone basked in the warm glow of partnership, however unequal. Some local businessmen started a Petroleum Club of their own. Fairbanks was on its way, or so everyone thought.

At the time, the most important thing seemed to be building a road from Fairbanks to the North Slope oilfields. The Fairbanks business establishment reasoned that, without a road, the city would be bypassed as a supply center. It would cost very little more to fly men and supplies to Prudhoe Bay from Anchorage than from Fairbanks, and Anchorage was already an industry supply center. Furthermore, it did not have ice fogs, a winter phenomenon which complicates flying into and out of Fairbanks. The oil companies were experimenting with less costly ways of getting heavy equipment to the Arctic: barges down the Mackenzie River to the Beaufort Sea or along the western coast of Alaska, through the Bering Strait, and down the Arctic Coast to the Sagavanirktok River. The Governor continued to talk about building a railroad but the Alaska Railroad was owned by the federal government and extending it would take a long time. Fairbanks businessmen and officials were convinced that a road north was the answer. Then the trucking industry entered the picture.

The truckers were unhappy because the airlines were making money hand over fist by flying freight to the North Slope. Interior Airways, which had had fourteen pilots in June 1968, had hired sixty by October and secured a lucrative contract with Pan American Petroleum, a subsidiary of Standard Oil of Indiana, to fly 200 trips to the Slope. To handle this, the airline had ordered new "stretch" Hercules cargo planes from Lockheed.

Immediately after the elections in November, Governor Hickel polled his new legislature to see how it would feel about a winter trail to the North Slope. Winter trails had been used in the Arctic before, particularly during the construction of the DEW Line. The legislators were generally amenable. The road was to cost about $350,000; the trucking industry had guaranteed that a certain tonnage would move over it during the coming winter; and there was talk of tolls to ease the taxpayers' burden. Almost everyone connected with the project should have known better than to build the winter road but almost no one objected before Governor Hickel made the final decision to go ahead with it, after conferring with representatives of the trucking industry.

The winter road, later christened by Governor Hickel's successor with glorious but unintentional irony the "Walter J. Hickel Highway," was a dreadful mistake. When christening it, Governor Miller said, "This impossible road shall be known by the name of the man whose courage, foresight, and faith in the Great Land gave Alaska what surely will become one of its greatest assets." Instead, the road left a swampy scar across the Alaskan wilderness; it was a costly but sloppy job of engineering and construction; and it did not do what it was supposed to do—get lots of heavy equipment to the North Slope quickly and cheaply.

Rather than piling snow on top of the tundra and so preserving its insulative qualities, the State Department of Highways (which ended up doing the job itself after first letting bids) gouged out a trail with bulldozers, scraping off the tundra that covers the frozen soil underneath. When spring came, the soil (which is usually protected by this vegetative mat and so remains frozen) thawed, and the Hickel Highway turned into a canal. This was an embarrassment to Hickel, who by then was Secretary of the Interior and working hard to upgrade his credentials as an environmentalist.

Nor did the road do what it was intended to do. Since construction crews had gotten a late start, the road was only open for about a month during the winter of 1968–69. During that time about seven and a half tons of equipment, most of it destined for North Slope oil rigs, moved up the road by truck. This was but a fraction of the tonnage flown into Sagwon, where the road ended, by airplane during that winter, in spite of a series of accidents which plagued the giant cargo aircraft.

And the road cost twice what Governor Hickel originally said it would cost. There were no tolls. Erosion was so bad in some places that the road had to be rerouted the following winter, adding to the expense.

The winter road was a monument to lack of planning, the absence of public policy, and greed. And, unlike Mayor Boucher's foray into the oil capitals of Texas, the road left a lasting mark, both across the physical face of Alaska and across its public image.

To many Outside observers, the winter road was ample proof that Alaskans could not be trusted with the last remaining wilderness.

It would be happy to note that the road's failure, which became apparent as soon as it thawed in the spring of 1969, ended the hysteria which had seized Fairbanks and much of Alaska, but it didn't. As the year passed, merchants feverishly built up their inventories and construction companies geared up in anticipation that construction would start soon on a huge pipeline from Prudhoe Bay to Valdez in southern Alaska. Three North Slope oil companies had joined together to construct the multimillion dollar project. It was to pass near Fairbanks, as planned, and that city would be the staging area for some of its construction.

In January 1969, six months after the initial discoveries at Prudhoe Bay had been confirmed, state Natural Resources Commissioner Thomas E. Kelly reclassified more than two million acres of tentatively approved state land on the Slope for competitive oil and gas leasing. Kelly maintained that the state could make much more money if the land was leased this way. His action had the effect of negating previous noncompetitive offers to lease which might have been granted earlier had it not been for Secretary Udall's land freeze, which Secretary Hickel had continued. Some Alaskans had made offers on the land Kelly reclassified, and had then sold them to oil companies at great profit after the ARCO discovery, while retaining an interest in any future oil royalties. The roar of outrage over Kelly's action could be heard throughout the state but nowhere was it louder than in Fairbanks, where, according to the *News-Miner*, about 150 residents stood to lose a lot of money in the reclassification process.

The thwarted leaseholders sputtered and fumed. Finally, in February, they set up a night meeting in Fairbanks with Kelly. Kelly's opponents did most of the talking. The Commissioner was "claim jumping," miner and former State Legislator Bill Waugaman cried, summoning up images out of Alaska's turbulent past. "It looks as though Tom Kelly is taking the land away from the people of Alaska and selling it to the big fellows," he shouted. Another speaker was Fairbanks oil lease broker Cliff Burglin, whose North Slope success story had made him something of a local hero. Burglin and a partner had paid $223 an acre for two blocks of land adjacent to Prudhoe Bay at a competitive lease sale in 1967, then turned around and sold their holdings to oil companies for $823 an acre plus a royalty. Said Burglin, "The state is not suffering for money." The basic argument, advanced by lease brokers, petroleum geologists, miners, and bankers alike was the "little man" theme: the small companies and investors will put more into Alaska than the big oil companies. They will do more exploration, hire more Alaskans, and take more chances. But they can't afford competitive leasing. Kelly did not change his mind. Later, Governor Miller announced that leases on about one-third of the reclassified land would be offered for sale on September 10. From then on, tension built in the state.

The summer of 1969 was a period of frantic activity on the Slope, as oil companies tried to gather geological information on which to bid intelligently on land adjacent to the Prudhoe Bay oilfield. Alaska law requires sealed bids which cannot be raised once they have been submitted, so it is impossible for one company to gamble on the interest of another. The companies already holding leases on the Slope had an obvious advantage and drilled around the clock to see what they could find. The regular work shift at Prudhoe that summer was twelve hours on, twelve hours off, for as long as a man could stand it. Meanwhile, seismic crews scoured the unleased land and petroleum experts pored over data from the United States Geological Survey, still the best source of information about oil in Alaska.

All this activity took its toll on the land. In their haste, the oil companies disregarded the peculiar Arctic environment in which they had to operate, tearing up the tundra as if it were the Texas plains. The environment fought back. After 1969, the companies found themselves with sloppy ruts or gullies instead of roads and expensive repairs that never seemed to do the trick. They finally turned to the Naval Arctic Research Laboratory in Barrow for advice. The laboratory, located at Point Barrow since the end of World War II, has compiled a wealth of information about Arctic engineering, based largely on ten years of careless oil exploration in nearby Pet Four. Today, thanks to the laboratory's advice, ARCO's Prudhoe Bay installations, the showplace of the North Slope, look more like a park than an oilfield.

The companies paid for their carelessness in 1969 in another way, too. Conservationists learned what was going on, were horrified, and skillfully orchestrated public opinion outside of Alaska against oil development there. By the end of the summer, the proposed pipeline was well on its way to being a *cause célèbre.*

As the sale date approached other tensions developed. There were fears that the sale might be illegal. (Two lawsuits were brought afterward, one by a former Attorney General of Alaska, who wanted to test the legality of the proceedings, and one by North Slope Natives, who claimed the state didn't own the land in the first place.) At the same time, the magic figure of $1 billion was beginning to creep into the public consciousness. It was in this atmosphere that the Twentieth Alaska Science Conference convened at the university in August.

Scarcely two weeks later, on the morning of September 10, the first sealed bid was opened before a standing room only crowd in Anchorage's Sydney Lawrence Auditorium. BP and Gulf together had offered $15.5 million for one block of land in the Colville delta. There was a gasp of astonishment from an audience that was still unused to thinking in millions of dollars. The bid on the second tract was higher; so was the one on the third. Ninety-seven million, seven hundred thousand dollars and a few minutes later, BP and Gulf controlled all six tracts in the delta. The sale was off to a thrilling start.

The highlight of the day was the bidding on Tract 57, the closest to Prudhoe Bay. The first bid opened was ARCO's, for $26 million. The second was made by the BP-Gulf combination and was for nearly twice as much, $47.2 million. Then Kelly opened bids by Continental, Sun Oil, and Cities Service for $36.6 million and the Hamilton Brothers group (which included Sun, Cities Service, and several others) for $36.8 million. BP still had the high bid. Next came the Mobil-Phillips-Standard of California bid and it was for $72.1 million. The audience cheered. Surely that was the winning bid. But, a few seconds later, Kelly announced that Amerada-Hess and Getty together had offered $72.3 million, or $28,233 an acre, for Tract 57. It was the high bid of the day.

The sale ended at 5:30 P.M. A plan was standing by to fly certified checks for 20 percent of the $900 million the state had collected to the Bank of America in San Francisco, where it could start earning interest immediately. It had been a spectacular day. The average price paid for North Slope leases was $2,000 an acre. BP and Gulf had spent more that day on four square miles along the Arctic Ocean than Alaska had made at all its previous lease sales. Relatively small companies like Louisiana Land and Exploration, Marathon Oil, Getty, and H. L. Hunt had spent $250 million on other land.

Outside the auditorium, a small group of Eskimos calling themselves Concerned Alaska Native Citizens and organized by Eskimo militant Charlie Edwardsen marched up and down all day carrying signs that read "$2 Billion Native Land Robbery" and "Bad Deal at Tom Kelly's Trading Post." Asked about the pickets by a local reporter, AFN official Don Wright replied that they did not have the support of the "official Native organization."

The lease sale was the psychological high point in Alaska's oil boom. Things started going downhill after September 10. Later that month, under pressure from the truckers, Governor Miller announced that he would reopen the winter road to the North Slope. It eventually was reopened, although only part of the way to Sagwon. But the winter road was completely unnecessary that winter. The oil companies cut back sharply on drilling as they waited for the pipeline to be built.

All through the long winter, Alaskans debated about what to do with their money. While the Legislative Council hired the

Brookings Institution, Governor Miller sought advice from a 24-man team from Stanford University and New York petroleum consultant Walter J. Levy. It is remarkable in a state whose citizens generally regard the Outside with a mixture of apprehension and derision how much Outside help they sought. Everyone talked about the best and wisest way to develop Alaska and almost everyone assumed that Alaska should be developed.

One man who questioned this assumption was an economist named Arlon Tussing, who was something of an iconoclast. "There is a philosophy of development for development's sake," he said at one of the Brookings seminars, "and almost everybody at this seminar shares that philosophy to one degree or another. But this is a set of attitudes peculiar to certain classes of people—businessmen, politicians, and upper civil servants, economists and the like—exactly the kinds of people invited to these seminars and interested in the questions to which it is addressed. The majority of Alaskans may not be for development for development's sake; most villages are not, nor are the oil workers on the Slope, nor fishermen, nor particularly . . . are most Alaska homesteaders. I even suspect that most of the 15- to 20-year-old children of the people here have little use for that philosophy." [86]

Apparently no one in the state or federal government, the Alaska business establishment, or the oil industry was listening to Tussing.

6

PRIVATE INDUSTRY'S MOST EXPENSIVE UNDERTAKING

"How long are we going to permit the private sector to decide these issues for us?"

Senator Gaylord Nelson,
September 1969

In June 1969, the Trans Alaska Pipeline System (TAPS), an unincorporated joint venture of three major oil companies, Atlantic-Richfield, British Petroleum, and Humble Oil, applied to the Interior Department for permission to construct a hot-oil pipeline across nearly 800 miles of public domain in Alaska.

As an Alaskan, Secretary of the Interior Walter J. Hickel was under great pressure to grant the construction permit promptly. The oil companies asked him to approve it by the first of July. Back in Alaska, where Secretary Hickel was building a base for future political adventures, or so he hoped, almost everyone wanted the pipeline built immediately.

However, he was also under pressure from conservationists not to allow the pipeline to be built promptly. The $900 million project posed innumerable questions—environmental, technical, and economic—to which no one had any answers. Secretary Hickel did not want to run afoul of conservationists so early in his career at the Interior Department.

The oil companies, perhaps believing in their own mystique, were supremely confident that Oil is omnipotent. Although they did not have a permit to build the pipeline yet, the consortium had already made a number of irrevocable decisions which now seem both arrogant and foolish.

The first was the decision to build a conventional buried pipeline. The oil in the pipeline had to be kept hot because cooling and reheating Prudhoe Bay crude turned it to tar. And the friction generated at pumping stations would heat the oil to nearly wellhead temperature—about 160°F.—anyway. But, instead of taking this heat and the presence of delicate permafrost along most of the proposed route into consideration and designing a pipeline to fit the situation, the companies simply decided to make a conventional pipeline fit Alaska. The initial feasibility studies for the pipeline consisted mostly of flying over Alaska in airplanes.

The selection of the route itself was arbitrary, despite a number of brief studies of alternatives commissioned in 1968 and early 1969 by the parent oil companies. In August 1968, less than a month after the original estimate of the North Slope's potential had been made, ARCO and Humble signed a contract with Pipe Line Technologists, Inc. of Houston to study the feasibility of a pipeline through Alaska to an ice-free port. TAPS appears never to have considered seriously an alternative route, although in December 1968, ARCO, Humble, and BP commissioned a second study which included a short section on an oil pipeline from Prudhoe Bay to Chicago, following the Mackenzie River in Canada. Another assessment of this route was being made at the same time, although the three North Slope giants were not directly involved in it. The Trans Mountain Oil Pipe Line Company of Vancouver, British Columbia, had commissioned the Bechtel Corporation to look into the feasibility of transporting Alaskan oil to a point in Canada where it could be fed into existing Canadian pipelines. All three studies concluded that a Canadian pipeline was economically feasible, despite its greater length than the Alaskan line and the presence of permafrost along portions of the routes in question.

TAPS opted for an Alaskan pipeline. In February 1969, the three companies announced that they planned to build it from Prudhoe Bay to Valdez, a small fishing village on Prince William Sound. The companies sought from the Interior Department permission to conduct geological and engineering investigations along the proposed route. This required a modification of the freeze on all transactions on public lands in Alaska. The Senate Interior Committee approved the modification and the Interior Department gave the companies permission to begin their investigations.

Another early decision had to do with the pipe itself. The companies had already decided that they would need 48-inch-diameter pipe in order to transport two million barrels of crude oil a day when the pipeline was in full operation. TAPS also decided that they needed the pipe by September, since they expected to get the necessary permit by July 1 and to start construction immediately. However, no American company made pipe that size. The only place to get it was Japan. At least two American companies offered to make special arrangements to produce the 48-inch pipe. U.S. Steel offered to convert an existing mill in Texas. Kaiser Industries offered to build a mill in Alaska. But since neither company could produce 48-inch pipe by September, the contract went to three Japanese companies—Sunimomo Metal Industries, Ltd., Nippon Steel Corporation, and Nippon Kokan Kabushiki Kaisha.

TAPS' rejection of the Kaiser offer particularly annoyed Secretary Hickel and other Alaskans who were interested in luring industry to Alaska. TAPS' choice of Japanese companies also irritated Representative John P. Saylor, the ranking Republican member of the House Interior Committee, for reasons which are unclear. Representative Saylor, whose Committee was directly involved with both the land claims and the pipeline, represents a coal-mining district in Pennsylvania. Although Bethlehem Steel is not a constituent, Representative Saylor had received sizable campaign contributions from officers of the Bethlehem mines, a subsidiary. At one point, when oilmen were explaining the TAPS project to members of the House Interior Committee, Representative Saylor asked why the companies had not ordered their pipe from Bethlehem Steel. The oilmen explained the need for 48-inch pipe, which by that time was beginning to arrive in Alaska from Japan. Representative Saylor was skeptical and added that TAPS had made its big mistake when it decided to buy Japanese pipe. Then, according to one of the oilmen present, the Congressman said, "You've just ruined yourselves as far as I'm concerned. I'm going to fight you every way I know how and the only way I know how is conservation."

All these crucial decisions were made before TAPS even asked for permission to build the pipeline.

In April 1969, still before TAPS had formally sought permission to build the pipeline, Secretary Hickel announced that he was setting up a departmental task force to oversee North Slope oil development, with Undersecretary Russell E. Train at its head. Train, who had been brought into the department to counteract conservationist opposition to Secretary Hickel, was a former head of the Conservation Foundation and later went on to become the first chairman of the Council on Environmental Quality. He and the Secretary never found working together easy. At the time, Secretary Hickel said his task force would work with the oil industry. A month later, under pressure from conservation groups, President Nixon expanded the task force to include a "conservation/industry ad hoc committee" and representatives of other government agencies. The President asked for a report on plans to protect the Arctic environment by September 15, 1969.

On June 6, TAPS filed an application for "an oil pipeline right-of-way together with two additional rights-of-way for ingress and egress to the primary right-of-way and eleven pumping plant sites for the construction of a 48-inch diameter oil pipeline system. . . . The primary purpose . . . is the construction, maintenance and operation of a 48-inch diameter pipeline and pumping stations for the transportation of liquid crude petroleum from the North Slope of Alaska to a marine terminal at Port Valdez." [87]

Immediately, Undersecretary Train sent TAPS chairman R. E. Dulaney a list of questions designed, according to Train, "to indicate the kind of questions to which satisfactory answers will be required before permits can be given for the use of public lands." It would be helpful if he could have Dulaney's answer within four weeks, Train wrote, "in our anticipation of industry's timetables." [88]

Dulaney lost no time. On June 19, he wrote that TAPS agreed "that our project should be constructed consistent with wise conservation and [we] reiterate our earlier observation that good pipeline design dictates design and construction procedures that will cause a minimum disturbance to the natural environment. Since we start from a common position, it is to be expected that the details of our proposed development will be consistent with sound conservation practices." [89] And he sent along a twenty-page document detailing the planning and research which had been done to date. Briefly, TAPS planned to bury the hot-oil pipeline, crossing all major rivers including the Yukon beneath the stream beds. There would be a system of block valves to shut off the flow of oil should a leak or break occur. There would be crews and equipment at various intervals to clean up the oil that did spill. TAPS had commissioned a variety of studies of ways to preserve or revegetate the tundra. And the companies were taking core samples along the proposed route to determine what kind of permafrost they would encounter. This was the first of many such exchanges between the companies and the Interior Department.

In the meantime, the Interior Department sent its own men to Alaska to reconnoiter. The Interior Department expedition traveled the length of the pipeline while Train and Dulaney were exchanging their letters. After the trip, the head of the expedition wrote a memorandum for Dr. William T. Pecora, then chief of the United States Geological Survey. "I do not think it is possible to bury most of the pipeline as anticipated by TAPS," he wrote on

June 20. "There appears to be a conflict between the proposal to bury all but 40 miles of the line and discussions with the University of Alaska people and Dr. [Max] Brewer [head of the Naval Arctic Research Laboratory]." [90]

Apparently, Dr. Pecora had reached that conclusion for himself. In May, he had toured the route of the proposed pipeline and talked to Brewer, an expert on Arctic engineering and permafrost. "Brewer indicated that there was not one chance in a hundred that any pipe would be buried north of the Brooks Range. We indicated that TAPS was considering burying pipe in the Sag delta and along the Sag River. Brewer said occasionally the river gets five miles wide and rechannels tremendously. He recommended the pipe not be buried in the Sag River." [91] Brewer and Pecora also talked about the haul road which TAPS wanted to build roughly adjacent to the pipeline in order to move equipment and supplies. "In construction, he [Brewer] felt that from Fairbanks north the pipe should follow the existing road [the Hickel Highway] and that the new road should be built only a few miles ahead of pipe construction," Pecora noted.[92] He also observed that the Hickel Highway "would take a long time to repair itself, if ever." And Pecora recorded his impression of the attitude taken by some oilmen toward the huge project. "A member of the ARCO group indicated that the oil must be taken out because of the responsibility to the people of the U.S. to make available to them this needed fuel and power source," Pecora wrote. "He indicated that there was no one there to catch fish in the rivers or to hunt anyway." [93]

TAPS chairman Dulaney's response to Train's questions had not supplied the kind of precise information the Interior Department needed. On June 27, Secretary Hickel wrote Dulaney: "We assure you that as soon as we are satisfied that you have met the requirements of law and regulation, that the interests of the Native peoples have been safeguarded, that environmental values will be adequately protected, that the responsible committees of Congress are in accord, and that consultation and coordination with other federal agencies and with the state of Alaska have been achieved, we will grant the necessary permit or permits as expeditiously as possible." [94]

In June, there was no National Environmental Policy Act. However, the company had to comply with the 1920 Mineral Leasing Act, which specified that a pipeline right-of-way should consist of the ground necessary for the pipe, in this case four feet, and twenty-five feet on either side of the pipe—a total of fifty-four feet. Yet TAPS was asking for a 100-foot right-of-way. Also, the land freeze was in effect and any modification of it would have to be approved by the Interior Committees of Congress in accordance with the promise the AFN and Senator Jackson had extracted from Secretary Hickel.

TAPS also planned to build a haul road to the North Slope near the pipeline from a point north of Fairbanks where the last public road ended. TAPS and the state of Alaska worked out a peculiar arrangement under which the company would build the road to state secondary road specifications and, when it was through with it, would turn the road over to the state for use as a state highway. Thus the people of Fairbanks and the Alaska trucking industry would get their road to the north, at no expense to the state.

The delay imposed upon TAPS by Secretary Hickel made Governor Miller nervous. On July 22, the state asked for a modification of the land freeze to allow a state highway to be built from Livengood, the furthest north one could get by road in the area, to the Yukon River, at the place where TAPS hoped to have its pipeline cross the swift river, a distance of about fifty-five miles. On July 29, Secretary Hickel asked the Senate Interior Committee to approve such a modification of the land freeze order. The Committee did and, on August 13, Secretary Hickel lifted the freeze for this purpose. TAPS promptly let the contract for construction to the Burgess Construction Co. of Fairbanks (which was soon to merge with the Houston Pipeline Co. of Houston, just one of many such mergers which followed the discovery of the

Prudhoe Bay oilfield). Work began promptly on the first segment of the Trans Alaska Pipeline project.

Five days after the September 10 lease sale in Anchorage, Undersecretary Train's task force submitted its preliminary report to the President. It was more of a status report on the application than a statement of how to develop the Alaskan oil discoveries with a minimum of damage to the environment. However, the task force had also drawn up a set of stipulations to govern the pipeline's construction. The task force's approach was distinctly that of putting the cart before the horse.

> Following the finalizing of the stipulations for the Trans Alaska Pipeline, the immediate focus of the task force will turn to the considerably broader problems weighing upon the proper administration of the public domain in Alaska—particularly those on the North Slope—involved with mineral and seismic exploration, drilling and oil and related mineral resource development. This effort will be a natural outgrowth and expansion of the present concentration on the pipeline right-of-way and a great number of the stipulations currently under development for the pipeline can be appropriately extended to the larger problems arising from future resource development.
>
> In our progress we have been guided by the view that oil development and environmental protection are not inconsistent. We have attempted in our proceedings to reach an equitable balance between the just concerns—and timetables . . . of industry, and the just concerns . . . and timetables . . . of the public.[95]

The task force also listed problems to which no solutions had been found by the September 15 deadline: permafrost, a source of gravel for insulating the pipeline from the ground, possible earthquakes along the route, the disposal of human waste, water pollution from oil spills or tanker discharges, and the effects on wildlife of so much human activity in what had been wilderness. Another problem was the Natives and their claims. And then there were several legal obstacles, including the size of the right-of-way

TAPS wanted. The task force could not decide how to give the oil companies the extra feet they claimed they needed. The report made it clear that the Interior Department was moving toward approval of the project on the theory that TAPS would provide the necessary technical information if prodded sufficiently. The report was kept secret until late fall.

But not everyone was as comfortable about the project as the task force. One afternoon in September, TAPS put on a slick slide show for Senator Jackson's Committee, at which the oilmen glibly assured the Senator and his colleagues that oil technology could cope with all the engineering problems the pipeline presented.

The oilmen exuded confidence. Since September 10, five more companies had joined TAPS: Mobil, Phillips, Union Oil of California, Amerada-Hess, and Home Oil. ARCO had already ordered three huge 120,000 dead weight ton supertankers to carry North Slope crude to its new refinery in Bellingham, Washington. And, as project manager George Hughes proudly told the Committee, the first of the Japanese pipe was due to arrive in Valdez any day.

Senator Jackson raised some questions about environmental aspects of the project. The Senator was being pushed by conservationists to question the pipeline closely. Conservationist opposition to TAPS had begun to take shape that summer, and by fall organizations like the Sierra Club and the Wilderness Society were urging members to write their Congressmen about it. But the only point at which Congress had any real control over the decision-making process was when Secretary Hickel asked the Interior Committees to approve a modification of the land freeze for the right-of-way. Conservationists knew the timetable for the project. Secretary Hickel was expected to approach Congress about lifting the freeze sometime that fall.

In the meantime, the Interior Department and TAPS set about removing the legal obstacles to the project. On September 19, the Native villages that claimed land over which the proposed pipeline would pass waived their claims to the right-of-way. On September

30, the Interior Department published the first of many sets of stipulations for construction of the pipeline. The thirty-four pages of stipulations were touted as the most rigid controls ever imposed upon a private construction project by the government. In fact, the burden of applying them rested largely upon Bureau of Land Management personnel in the field, who would be working alongside the construction crews supervising the project. The idea was that these supervisors would make sure TAPS and its contractors obeyed the stipulations and would close down construction when they did not or when a question arose which could not be answered in the field or to protect fish and wildlife at crucial times like nesting, spawning, or migration. In short, it would take inspectors with an unusual amount of guts to do the job properly.

The Secretary's idea was to give TAPS permits for portions of the pipeline as the companies supplied the necessary information, thus allowing construction to start on one part of the project before the whole project had been approved. And he was prepared to approve construction of the haul road, which required its own entirely separate right-of-way, immediately, on the grounds that roads had been built successfully in the Arctic for years. The road was really part of the total project, although Alaskans and oilmen argued it wasn't. Moreover, approving construction of the road immediately solved a pressing political problem for the Secretary—how to remain popular in Alaska while denying permission to the oil companies to build the pipeline until they had made at least a pretence of doing their homework.

Accordingly, on October 1, Secretary Hickel asked Congress to approve his raising the freeze for the entire project. The House Interior Committee seemed willing to give him their approval but, unable to round up a quorum for this action, put off any decision until the last week in October. In the meantime, the House Indian Affairs Subcommittee was to make a tour of Native villages, which included a flight over the route of the proposed pipeline. The Senate Committee was more skeptical. Senator Jackson said he would have to hold some hearings on the modification in order to get the information his Committee needed before making a decision. It would also give the conservationists an opportunity to testify against the project, thus getting them off Senator Jackson's back. The Senator was said to be toying with the idea of retaining oversight of the huge project for his Committee.

In mid-October, Undersecretary Train appeared before the Senate Committee to argue for lifting the freeze. "Lifting the freeze by itself allows no one the right to encroach upon, or proceed with, construction of any part of the pipeline route," he said. "It is simply a necessary procedural step that allows future action on the actual permit." Secretary Hickel would not let construction start until he was sure the companies' plans were safe, Train said. But he made it clear that the Secretary wanted the project to start as soon as possible. The record does not show that Train, the conservationist, had any trouble with these comments.

But Senator Jackson wondered if it wouldn't be better to "deal with the known and defer to the unknown" as far as construction permits were concerned. Train said TAPS understood the Department's position. It was not until much later that the Department admitted that TAPS was unwilling to accept piecemeal permits. The Department and TAPS were operating within a "time frame" created by the companies' investments and decisions, Train said.

"How long are we going to permit the private sector to decide these issues for us?" asked Senator Gaylord Nelson, D-Wis., angrily.

"Much as we need oil from the North Slope and would like to accommodate these people who've invested billions of dollars up there," said Senator Metcalf, "we should proceed cautiously."

Train told the Committee that TAPS hoped to start work on the Valdez-to-Fairbanks segment of the pipeline in March 1970. But he hinted at a problem there: particularly unstable permafrost in the Copper River Valley north of the coastal mountains. The soil there was silty and contained a great deal of ice, which could be melted by the heat from a buried pipeline. If the permafrost supporting the pipeline were to melt, then the pipe would bend dangerously, perhaps even break. TAPS insisted it could bury the pipe there but could not tell the Geological Survey how it could do it safely.

When the Undersecretary appeared before the House Interior Committee, he was taken completely by surprise. The two ranking members, just back from their tour of Alaska and the pipeline route, were furious because they had seen "denuded" hilltops, a swath cut in the wilderness for "tens of miles" and all sorts of "drilling" activities. "The things the committee saw were enough to raise the hackles on the back of even a bald head," remarked Representative Saylor. But what disturbed the Congressmen more than the desecration of the wilderness was the fact that it had been desecrated without their permission. "After seeing all the work that's been done up there," continued Representative Saylor, "it's rather like coming to us after the horse has been stolen. The only thing left is to lay the pipe. The rest's been done." Train was taken aback and later sent the Congressmen a message that the "denuded" hilltops had been gravel borrow pits for construction of the road to the Yukon and the swath through the wilderness had been cut to allow TAPS to collect core samples of the soil along the pipeline route, both activities they had approved.

On October 23, Senator Jackson sent Secretary Hickel a list of questions to which he said he needed answers before his Committee could approve any modification of the land freeze for the project. These questions dealt with almost every phase of the pipeline, from permafrost to the whereabouts of the task force's report to the President (still a secret); from the delegation of federal oversight authority to the rights of the Native people. When he learned about Senator Jackson's questions, Chairman Aspinall decided that the House Interior Committee would wait until the Secretary had answered them before approving the freeze modification.

It took Secretary Hickel nearly a month to answer Senator Jackson's questions. Asked if granting a permit for the pipeline while Congress was drafting a claims settlement was in the best interests of the Natives, the Secretary replied that "the granting of a pipeline right-of-way will not be harmful to the best interests of Alaska Natives." Asked if the Natives favored the pipeline, Secretary Hickel hedged and replied, "No Native village or association has reacted unfavorably to the pipeline project." Asked how large a right-of-way TAPS was seeking, Secretary Hickel wrote: "Section 28 of the Mineral Leasing Act . . . authorized the granting of rights-of-way for oil or natural gas pipelines to the extent of the ground occupied by such pipeline and 25 feet," which was true as far as it went. He did not mention the extra land TAPS wanted. Asked if the Interior Department had any "departmental policy" for opening up the wilderness north of the Yukon River, the Secretary replied, "No." Asked whether any of the land through which the pipeline would pass was being considered for national parks, wilderness areas, or wildlife refuges, the Secretary replied evasively, "If it is found that pipeline construction would seriously impair public values, the rights-of-way will be adjusted."

The Secretary said the major public benefits of the pipeline would be to supply badly needed oil to the West Coast and jobs to Alaskans. The major public risk was a disastrous oil spill, he said. Asked if these benefits outweighed the risk, the Secretary's answer was curious: "The department's communication with and testimony before appropriate committees of Congress concerned the lifting of the land freeze for the pipeline right-of-way and not the granting of the actual permit."

The Secretary was straightforward about TAPS' engineering problems, though. He told Senator Jackson TAPS had not solved these to his satisfaction.

Back in Alaska, Governor Miller was getting more and more nervous about the pipeline. This was at the height of the anti-Native feeling and emotions were stretched almost to the breaking point. Governor Miller began to see plots everywhere. The Governor was not the only person in Alaska who was confused about the status of the project. While Secretary Hickel repeatedly said publicly that he needed more information and plans from

TAPS before he could approve the project, many Alaskans just could not believe that the oil companies themselves were responsible for the delay. Instead they blamed the Secretary, Congress, national conservation groups, or the Natives.

In early December, Secretary Hickel sent a second letter to Senator Jackson, in which he reiterated that his department would do nothing until TAPS had solved its problems with permafrost. "We are confident that the state of technology is capable of solving these problems," he wrote. "However, at the present time TAPS has not yet provided us the necessary answers and assurances which would permit us to move ahead." [96] So, on December 11, the Committee notified the Secretary that it had no objection to raising the freeze for the pipeline as it did not seem to impede settlement of the land claims. However, Senator Jackson made it clear that his Committee was worried about the environmental effects of the pipeline. He also let the Secretary know that the Committee had considered, and rejected, the option of evaluating the pipeline itself. It took expertise the Committee did not have, the Senator concluded. Senator Jackson ended his letter with mention of the National Environmental Policy Act (NEPA), then agreed upon in its final form by a Senate-House conference committee and about to be submitted to both houses for a last vote. It "provides a Congressional statement of national goals, and of basic guidelines and policies which you and the representatives of your department may wish to consult," wrote Senator Jackson. It also provided, in Section 102 (c) (2), for a public statement of the environmental impact of any federally financed project or any project on federal lands, including an evaluation of alternatives to it. A few days later, following the Senate's lead, the House Committee also notified Secretary Hickel that it did not object to his lifting the freeze for the pipeline. Representative Saylor, the friend of Bethlehem Steel, alone voted against allowing the modification. Secretary Hickel did not immediately grant the permit, although he modified the land freeze to allow himself to do so. However, the BLM started the process of classifying five million acres along the proposed pipeline route as a transportation corridor. And TAPS refiled its application for a permit, this time requesting a legal 54-foot right-of-way and temporary Special Land Use Permits for additional footage up to 146 feet in some places along the route.

What Secretary Hickel planned to do first was issue a permit for construction of the 390-mile haul road. In January, TAPS issued letters of intent to contractors for the entire road from the Yukon to Prudhoe Bay. The bulk of the work was to go to Burgess, which had built the Livengood Road, and the Green Construction Company of Fairbanks. Together these firms were to build 260 miles of the road. The other 130 miles were divided between three other companies. These companies, anticipating firm contracts soon, started marshaling their equipment in Fairbanks. Their plan was to move the heavy construction machinery up the old Hickel Highway, which had been partially reopened, to convenient staging points along the haul road's tentative route. The companies wanted to get their equipment to these locations before the actual right-of-way permit was issued, so that, when it did come through, construction could begin immediately. Time was important. The ice road was expected to thaw sometime between the first and the fifteenth of April and the equipment had to get to the staging areas before then. It was particularly important because the ice bridge on the Hickel Highway was the only way to get the machinery across the Yukon River.

The state, which had opened the winter road halfway, as far as the Native village of Bettles, started negotiating with some of the contractors to open the road beyond that point. At the same time, representatives of assorted Chambers of Commerce, alarmed by reports that flotillas of Outside barges were mobilizing in Seattle and on the Mackenzie River to carry freight to the North Slope that summer, descended on Juneau to press for reopening the Hickel Highway immediately. In March, the legislature appropriated additional funds. Soon hundreds of pieces of heavy equipment started up the road. About the same time, the BLM authorized TAPS to go ahead with the centerline survey for its own, completely new, road. A team of surveyors started north. By March 10, the road was two-thirds open and the surveyors had gotten well beyond the Yukon.

The situation was this: a new winter road, for much of the earlier Hickel Highway was now useless, was being carved across the wilderness to allow tons of heavy equipment to move into place so that a third road, this one permanent, could be built. All this activity put tremendous pressure upon Secretary Hickel to issue a permit for the haul road immediately, since if the equipment had to remain idle up north all summer, the companies which owned it would face staggering losses. And once the ice bridge over the Yukon had melted there was no way to get the machinery back to Fairbanks. Secretary Hickel was on the verge of issuing the permit for the road when both Natives and conservationists asked a federal district court in Washington, D.C. to prevent him from doing so.

In January 1970, after TAPS had named the contractors for the haul road, some of the villages which had signed a waiver of their claims to the pipeline right-of-way the previous fall wired Secretary Hickel asking him not to allow construction to start because TAPS had not chosen any Native contractors and had picked contractors who were not likely to give jobs to Natives living along the route of the pipeline. They said TAPS had promised them jobs. TAPS said it did not know what the villagers were talking about. With the aid of attorneys from the Alaska Legal Service, an OEO-funded organization, the villages of Bettles and Allakaket sued TAPS for breach of contract. Then, on March 9, before anything had happened in the case, five villages asked the federal district court in Washington to stop Secretary Hickel from issuing the construction permit on the grounds that they claimed the land over which the pipeline and the road would pass. For this suit, the villagers retained three young lawyers from the prestigious Washington law firm of Arnold and Porter.

On April 1, Judge George L. Hart, Jr., a jurist who was to learn more about Alaska in the next few years than he probably cared to know, sat puzzling over a rather unclear map of the state while the attorneys argued about land claims, a subject no judge in his right mind would touch if he could avoid it. Judge Hart avoided it, sensing that there was something else at the heart of the suit. He urged the Natives' attorneys and the Justice Department "to get busy and take the matter out of the courts entirely." Then he enjoined the Interior Department from issuing a construction permit across 19.8 miles of the route, land claimed by the sixty-six residents of Stevens Village, nestled in a bend in the Yukon not far from where TAPS proposed to cross the river. He said he would reconsider his temporary restraining order in ten days. "I hope in the interim you gentlemen will get the matter resolved," Judge Hart said to the attorneys. They all nodded in agreement.

However, in the meantime, three conservation groups—the Wilderness Society, the Friends of the Earth, and the Environmental Defense Fund—on March 26 sued the Interior Department in the same district court and asked it to halt the TAPS project on the ground that it violated both the 1920 Mineral Leasing Act and the new National Environmental Policy Act. After a number of preliminary moves, including assurances that no permit would be issued in the interim, the court scheduled a hearing in this case for April 13.

In Alaska, Governor Miller decided to take matters into his own hands. On the basis of an 1866 statute which grants states rights-of-way over public lands "not reserved for public use," the Governor "authorized" the haul road himself. While Secretary Hickel professed "shock" at the Governor's course of action, it is clear that the Interior Department had suggested it to the Governor in the first place. In a March 20 environmental impact statement on the haul road, still secret in early April, the Department said: "The alternative, therefore, is not between a road or no road, nor between a road now and a road three years from now, rather it is between a federally-controlled right-of-way and one that goes forward under state jurisdiction." [97]

The Governor's action was greeted with jubilation in Alaska

but not in Houston, the headquarters of TAPS. On April 7 and 8, TAPS and its contractors huddled to decide what to do. They opted to wait for a permit from the federal government. Governor Miller was alarmed. He said he couldn't understand why TAPS didn't start work immediately "unless TAPS is convinced there will be no pipeline ever." And he flew off to Texas to talk to the TAPS officials on April 12 and 13.

During all this political and judicial maneuvering, several informal groups met in Alaska to see what they could do about the delays. One such gathering of nearly 200 businessmen, contractors, and labor leaders met in Fairbanks on April 11, and collected $1,400 by passing the hat even before they had organized themselves. The money was to finance lobbying activities but no one knew whom they ought to be lobbying. The tone of the Fairbanks meeting was hysterical. One man said, "We'll have so many unemployed walking the streets, Fairbanks won't be safe. We must do something about that." [98] Others were angry with the Natives whose suit had led to the first temporary restraining order against the TAPS project. In Anchorage the same day, Senator Stevens urged a similar group to send a delegation to tiny Stevens Village to ask them to drop their suit. A group of Fairbanks businessmen eventually did fly to Stevens Village to try to coerce the Natives to do just this. The businessmen suggested that the villagers might not get things they wanted, like a new airstrip, if they were uncooperative. The villagers refused to drop their suit.

On Monday, April 13, attorneys for the conservation groups and the federal government, plus a bevy of oilmen and their lawyers, a huge delegation from the Interior Department, and some lawyers for the Alaska Natives gathered in Judge Hart's courtroom. The first speaker was James Moorman, a young lawyer who had left the government only a short time before to join a new public service law firm called the Center for Law and Social Policy. After a brief discussion of the discrepancy between the size of the right-of-way TAPS wanted and the size allowed by the Mineral Leasing Act, Moorman launched into the heart of his complaint.

"A purported environmental statement" had come to light, he told the judge, which mentioned none of the environmental hazards the project posed. In fact, it did not even mention a pipeline. He was referring to the Interior Department's still-secret eight-page impact statement of March 20. "It's an attempt to paper over the problems by ignoring them," said Moorman. And he started to tell Judge Hart about the dangers to the environment. The judge cut him off quickly. "The question before me is not whether they're doing this stupidly but whether they're doing it legally," he said.

Then Herbert Pittle, a small white-haired lawyer from the Justice Department, had his turn. Pittle did not seem to know much about the project, but then the Interior Department had given him very little to go on. Pittle told the judge that the Interior Department and TAPS had no firm plans for the pipeline. He admitted that there was no impact statement for the pipeline, only for the road, but agreed with the judge that the only reason for the road was the pipeline. Judge Hart then asked about the Mineral Leasing Act. Pittle told him that TAPS really had three applications before the BLM, one for a 54-foot right-of-way for the pipeline, one for a 200-foot right-of-way for the haul road, and a special request for additional footage alongside the pipeline. "What was the sense of Congress making this law if you can give them 10 miles by giving them rights-of-way alongside one another?" asked Judge Hart. "You can't violate the law just by spending a billion dollars to do it."

Then Pittle admitted that the Department was about to issue the permit for the road. Judge Hart asked Moorman what relief he sought. The lawyer replied that since the Interior Department was about to give its permission for the road to be built, his clients would like a preliminary injunction until the case could be tried on its merits, and two weeks' notice of the Department's intent to issue the permit. "Fair enough," said the judge.

When Pittle protested, the judge told him, "The road and the pipeline are all one thing. The method you propose would violate the Environmental Policy Act of 1969 and the Mineral Leasing

Act." He issued a temporary injunction against the TAPS project.

Several important documents came to light during the hearing. One was a March 25 memorandum from U.S. Geological Survey head William T. Pecora to Secretary Hickel. Pecora wrote: "The collective judgment of the work group [at Menlo Park] is that TAPS has not demonstrated acceptable fundamental design criteria for below-ground construction in permafrost of a hot-oil pipeline that would be reasonably safe from failure." Since it was difficult to justify a buried pipeline, Pecora continued, TAPS should look into above-ground construction, perhaps even alternative routes. And he concluded: "Disclosure of this conclusion may be required soon in view of public expectancy of results of study." [99]

Up to that time, TAPS had insisted that about 750 miles of the 789-mile pipeline could be buried. Pecora elaborated on his memorandum two weeks later when a delegation of Alaskans came to see him about the pipeline. He told them that it would be unsafe to bury the pipe in permafrost for anywhere between 90 and 95 percent of proposed route. He described his meetings with TAPS engineers, who had apparently conceded that there was a fifty-fifty chance of the pipe breaking. "Hell, I wouldn't put up anything under those circumstances," said Pecora.

Alaskans were generally dismayed by the injunction against the pipeline but Secretary Hickel continued to tantalize them, as at an Earth Day rally at the University of Alaska on April 23, when he abandoned his text and said, "I am announcing tonight that I will issue the permit for the pipeline right-of-way." [100] But not then.

In the meantime, Governor Miller proposed to the parent companies of TAPS that the state build the road and the oil companies pay for it. He said the presidents of ARCO, BP, Humble, and Standard Oil of Ohio, which was then in the process of merging with BP, had tentatively agreed to this.

The oil companies were beginning to grasp the situation in Alaska, something they had woefully miscalculated in 1968 and 1969. The two American companies originally involved in TAPS—

ARCO and Humble Oil—simply couldn't believe that there would be any problem getting permission to build a pipeline across public lands. In fact, they assured their foreign partner, BP, that the industry-oriented Interior Department had never before turned down such a request, which was true. However, they ignored the peculiar situation in Alaska, where title to most of the land was unclear and there was a freeze on all public land transactions.

Curiously, only BP had an Alaskan expert on their staff. BP Alaska had hired Hugh Gallagher, a former administrative assistant to the late Senator Bartlett, to represent them in Washington. Gallagher, who had great affection for the Natives and was a close personal friend of one of the most militant of the Eskimo leaders, Charlie Edwardsen, tried unsuccessfully to convince the company that it ought to involve itself in the land claims settlement. However, it was not until BP was approached by one of the lawyers for the Alaska Federation of Natives that Gallagher had any success. In the spring of 1970, while the Senate Interior Committee was still trying to draft a settlement bill, former Attorney General Ramsey Clark, now actively lobbying for the Natives, became worried about the House Interior Committee, which both the Natives and the state had previously ignored. Representative Haley, the Chairman of the Indian Affairs Subcommittee, which was nominally handling the bill (in reality, the Chairman of the full Committee, Representative Aspinall, was in charge of it), had just refused to talk to Clark about the Natives' case. Clark decided that the Natives would need help with the House Committee and that the key to it and its Chairman was the oil industry. Accordingly, Clark approached Glen E. Taylor, who had been an acting Assistant Attorney General for Land and Natural Resources during Clark's tenure in the Justice Department, and who was now Washington counsel for BP. Taylor set up a meeting for Clark with Gallagher and the president of BP Alaska, Frank Rickwood, to discuss the claims. As a result of that meeting, BP agreed to help lobby for a claims bill the Natives wanted and to try to persuade the other companies in TAPS to do likewise.

However, nothing happened immediately because TAPS was reorganizing. That fall the ponderous joint venture of which no one company was clearly in control became a new tightly-knit Delaware corporation, the Alyeska Pipeline Service Company, Inc. The same companies were involved, with ARCO, Humble, and BP holding 80 percent of the stock. One of the striking things about TAPS had been its inefficiency. Sometimes the major companies in the venture seemed to be operating at cross purposes. In fact, they may have been. BP had every reason to want the pipeline built as quickly as possible. For one thing, BP held about 60 percent of the proven reserves at Prudhoe Bay. Furthermore, its merger agreement with Standard Oil of Ohio (SOHIO) stipulated that the British company's holdings in SOHIO would increase as the North Slope oil came on line. In fact, BP's ultimate holdings in the American company would be determined by its throughput of Alaskan oil as of January 1, 1978. In order to acquire more than 25 percent of SOHIO's common stock, BP would have to move at least 200,000 barrels of its Alaskan oil through the pipeline every day by then. On the other hand, Humble Oil seemed to be dragging its feet. It is possible that Humble and its parent company, then Standard of New Jersey (now the Exxon Corporation), did not welcome the kind of domestic competition the BP-SOHIO combination would provide. At any rate, Interior Department officials who had to deal with TAPS and the parent companies regularly during this period generally agree that Humble was the least cooperative of the companies involved.

By mid-1970, plans for construction of the pipeline were stymied. There was an injunction against its construction, based partly upon environmental considerations. But the oil companies were beginning to wake up to the realization that the real stumbling block was the Native land claims. There could never be a Trans Alaska Pipeline until the question of who owned Alaska was resolved.

7
A LAND SETTLEMENT

"Take our land, take our life."
AFN slogan, 1971

Late on the afternoon of April 14, 1970, Senator Jackson stepped out of a conference room in the bowels of the Capitol building and handed waiting reporters a summary outline of the land claims legislation upon which his Senate Interior Committee had tentatively agreed. There were few surprises on Senator Jackson's summary sheet, since most of the bill's provisions had already been leaked to reporters despite the Senator's attempts to keep the deliberations a secret. The Committee had decided to give the Natives $500 million in federal appropriations over the next twelve years, $500 million from a two percent royalty on mineral revenues over an indefinite period, seven and a half million acres of land, and two statewide corporations to manage the settlement. Senator Jackson's summary gave no hint of what were to become controversial elements in the bill: a section requiring competitive mineral leasing in most of Alaska, and restrictions on private entries into public lands for the next five years.

After Senator Jackson's anger over leaks had led to the

discarding of the 1969 Alaskan "compromise," the Committee had started from scratch. Senator Jackson had the staff draft a new bill, which was almost complete when the Committee met for the first time in 1970, in late February. The staff proposed a federal payment to the Natives of $500 million plus a two percent royalty on all mineral leases for twenty-five years, and three townships of land for each village, provided the land was contiguous to the village. In addition, a statewide corporation would be given other land, like forests or mineral land, so that the revenues from it would be used for the benefit of all Natives. While the villages were to choose any land that was not actually patented to the state, the corporation was limited to land which had neither been patented nor tentatively approved for patent to the state. However, the Natives got to select their land before the state could make any more selections of its own. The Committee took this staff bill and refined it, producing in the end a piece of legislation that pleased almost no one in Alaska.

First, the Senators decided that the 40 million acres proposed by the Natives in 1969 were too much. Most of the members of the Committee were westerners, accustomed to thinking in terms of large acreage. But they balked at 40 million. Others, following the lead of the Federal Field Committee in its 1969 report, were concerned about creating large racial enclaves. Finally, after two months of wrangling, the Committee settled on seven and a half million acres, far less than the three townships per village which the staff had proposed, provided the Natives were allowed to use other public lands for hunting, fishing, and berrying. The villages were to get full title, that is, mineral rights as well as surface rights, to only four and a half million acres. A statewide corporation would get surface rights only to the other three million. Although the amount of land was later increased to ten million acres, it was far less than the Natives considered acceptable. "Ours is a land claim, reject Senate bill, we wish land," the village council of Unalakleet telegraphed their Senators in early May when the details of the

Senate bill finally reached them. Most other villages felt the same way.

Neither did the bill please the state. The Committee decided that the state should contribute to the settlement by matching the $500 million federal appropriation with a two percent royalty, amounting to $500 million, on revenues from all state and federal lands. Of the $500 million, $497 million was to come from money that would otherwise have gone into the state's pocket—state oil and gas lease royalties, rents and bonuses, and Alaska's 90 percent share of federal mineral revenues. The Committee's rationale was this: the Congress had been generous with Alaska, giving her an unprecedented percentage of federal mineral revenues plus an enormous land grant, and, therefore, Alaska could be similarly generous with her Native citizens. Furthermore, the pending Native claims were a cloud on title to all land in Alaska. Clearing title was costing the federal government a good deal of money, but the beneficiary was Alaska, not the federal government. The Committee wanted to make Alaska pay for this.

The news that the royalty would come mostly from state rather than federal funds brought new cries of anguish from Alaska's establishment, which had hoped the Committee would buy their Attorney General's argument that since the Statehood Act had been approved by Alaskan voters, it could not be altered without their permission. But the Committee decided that the two percent royalty did not violate the Statehood Act, and to be sure that the state did not challenge this part of the settlement in court, added a provision that should Alaska do so, its land selections would be suspended as long as the litigation lasted.

Not all members of the Committee liked the royalty either. Even after this revenue sharing provision had been agreed upon tentatively in early March, Senator Mark Hatfield, R-Ore., proposed letting the state match some portion of the federal $500 million directly instead of with a royalty. This would have required special state legislation. Senator Hatfield and others were nervous about precedent. Would other Indian tribes in other states seek a similar royalty? they wondered. The Committee also considered, but rejected, a provision to make the royalty apply retroactively to the $900 million the state had collected at the lease sale the previous year.

Even the $500 million in federal funds was questioned. Senator Anderson of New Mexico repeatedly tried to cut it to $300 million.

However, the most controversial issue as far as the Committee was concerned was one which did not involve the Alaska Natives directly—federal mineral leasing policy in Alaska. Competitive leasing was required on all federal land where oil and gas were known to exist. However, Senator Jackson was interested in changing the entire federal mineral leasing procedure and he saw the Native claims bill as a means of changing the system in Alaska and thus getting a foot in the door. Competitive mineral leasing made little or no difference as far as the Natives' two percent royalty was concerned but it did make opponents out of legislators representing petroleum interests and oil-minded Alaskans. The competitive leasing provision, Section 17, was included in the legislation the staff drafted in February. It was also in early printed drafts of the final bill. But pressure was building to prevent the legislation from getting out of committee in that form. Max Barash, lobbyist for Canadian oil companies, who would lose their cheap options to lease about 20 million Alaskan acres if the bill passed in that form, went to work. So did Senator Clifford P. Hansen, R-Wyo., a staunch friend of oil and mining interests and a member of the Committee. Telegrams and letters poured into the Committee members' offices from industry representatives. The showdown came on May 13, when the Committee met in closed session and Senator Hansen led a successful move to delete Section 17. The vote was ten to seven. Senator Stevens voted for its deletion, Senator Gravel against it. Afterward, lobbyist Barash was lavish with his praise. One Senator who had voted to delete Section 17 received the following message: "He [Barash] called on May 14 to thank you and tell you that you had served your country and your state well."

Senator Jackson did not give up on competitive leasing. He later reintroduced Section 17 on the Senate floor as an amendment to the claims bill and this time it was accepted. Senator Jackson also wanted a provision in the bill that, if Pet Four was ever opened to oil and gas leasing, it would have to be leased competitively, but he was unable to get the Committee to agree to it.

Some members of the Committee feared a land rush in Alaska once the claims had been settled. They argued that it was not enough merely to prevent entries into the public lands around the villages until the Natives had selected their land. Something had to be done about the rest of Alaska too. So the Committee decided to withdraw all public lands in the state from private entry, including the staking of mining claims, for five years. This became known in Alaska as the Five-Year Freeze and was generally disliked.

The Committee also put a lid on attorneys' fees the federal government would pay. The Committee was not so worried about Arthur Goldberg and his associates as it was about the numerous Alaskan lawyers who had regularly represented Native groups or villages over the years. The Committee decided that no more than $1 million should be authorized for lawyers' fees and that each claim by an attorney would have to be approved by the chief commissioner of the Court of Claims before it was paid. The provision was popular in Alaska with almost everyone except the attorneys. One complained to Senator Stevens that his partnership had broken up over the financial strain created by his representation without pay of Native groups. "I never expected to become wealthy from undertaking this work," he wired, "but must you impoverish me?"

Thus the Committee had reduced the amount of land the staff version would have given the Natives, it had increased greatly the financial participation of the state of Alaska, and it had deleted provisions unfavorable to the oil companies.

Before Senator Jackson announced the tentative agreement his Committee had reached in mid-April, most of the lobbying had been done by groups like the American Mining Congress and the

Alaska Miners Association. The major oil companies were still uninterested. Barash's real work came later. Meanwhile the conservationists were preparing for a direct onslaught against the pipeline in federal court.

Once the AFN learned what the bill contained for the Natives, there was widespread dissatisfaction with it. The AFN lobbyists, among them Eben Hopson and John Borbridge, Jr., officials of the AFN, and Ramsey Clark, their lawyer, tried to change the bill but scored only minor successes. Before reporting out the bill in May, the Committee did agree to raise the amount of land to 10 million acres, although for various reasons there was no guarantee that it would work out to be that much, and to give the North Slope people surface rights to an additional 500,000 acres around their villages. Both Alaskan Senators urged the Natives to accept the bill with these relatively minor changes and try to win substantive changes in the House of Representatives. The Natives, however, knew that the House Committee was less inclined to be generous than the Senate panel. After numerous meetings, both Senators agreed to support amendments on the floor that would give the Natives no less than 10 million acres and the North Slope people mineral as well as surface rights to their additional 500,000 acres. But the Natives still wanted the 10 million raised to 40 million and neither Alaskan Senator was willing to go along with that. The two Alaskans said the AFN amendment would not pass the Senate and could lead to attempts by conservative members of the Committee like Senators Allott and Fannin to gut the settlement by reducing the federal appropriation and eliminating the royalty. Senator Gravel eventually did cosponsor this 40-million-acre proposal once a third Senator, Fred B. Harris, D-Okla., introduced it on the Senate floor, and Senator Stevens reluctantly voted for it.

The AFN had finally found a champion in Senator Harris, a populist politician whose Comanche wife, LaDonna, had been a fervent supporter of the Alaska Natives all along. Senator Harris agreed to introduce all the amendments for which the AFN could not find Committee sponsors.

On July 14, after 103 years of waffling by the United States government, the Senate finally began debating the claims. Despite fears that too many amendments would be offered and that opponents of the settlement would gut it, the bill passed with remarkably little discussion. The most controversial section, predictably, turned out to be the Jackson amendment on competitive leasing. The Senate approved this by a margin of twelve votes, thus reversing the Committee's action. But the AFN lost every attempt to liberalize the bill that did not have the backing of the Alaskan Senators or the Committee. The Natives did not get the 40 million acres they wanted, but they succeeded in increasing the seven and a half million acres to ten million and in obtaining surface and mineral rights to an additional 500,000 acres for the North Slope people alone. The federal government was to give the Natives $500 million and they were to get another $500 million in mineral revenues, to be collected mostly from money that would otherwise go to the state of Alaska.

Eight Senators ended up voting against the settlement. They did so for a variety of reasons. Senator Kennedy voted against it because he said the Natives wanted more land and ought to get it. Senator John J. Williams, R-Del., voted against it because he opposed the royalty. Senator Hansen objected to competitive leasing. But most Senators, unable to sort out the complexities of the legislation for themselves, took the Committee's advice and approved the bill.

The Natives were not happy, although Ramsey Clark wrote a letter to Senator Stevens thanking him for his assistance, in which he said, "I think it is wonderful that the Senate was able to pass a bill that the Natives will be able to accept." Communications between the AFN and their attorneys were breaking down. There were many Natives who were not willing to accept the Senate bill. Many also felt that, while they had hired the best available legal talent, they had little to show for it. Goldberg was off running for Governor of New York. And Clark's efforts had gotten them a bill their people did not like from the Senate and nothing whatsoever as yet from the House.

It was at this point that attorney Clark decided to approach the oil companies for help (see Chapter 6). Unfortunately, this approach was made too late in the year to do much good during that session of Congress.

By the time the Senate passed a claims settlement bill, there were only three weeks left until Congress took its summer recess. Little could be done in the House Committee in that amount of time. Worse still, 1970 was an election year and Congressmen had to think about getting reelected. The Alaskan member of the House Interior Committee, Representative Howard Pollock, was in Alaska most of the time that year campaigning for the Republican nomination for Governor, having decided to give up his Congressional seat. (He lost his bid for Governor during Congress' summer vacation.) It was an ill-concealed secret that the older members of the House Committee resented Representative Pollock's frequent and prolonged absences and felt he had not been doing his share of the work.

On Monday, after the Senate approved a claims bill, Representative Haley's Indian Affairs Subcommittee met in the Longworth Building. When he left that meeting, the elderly Chairman, who had felt ignored by white Alaskans during a tour of the state, said it might be months before his subcommittee could consider the land claims. He said there were many "Indian" bills with higher priorities than this one. However, the subcommittee set aside four days in late September to mark up a bill.

In the month before the subcommittee met again on the land claims, the various groups with interests at stake got to work. First there were the oil companies. During this time, the Trans Alaska Pipeline System completed its transformation into the Alyeska Pipeline Service Company. The new corporation was headed by Edward Patton, formerly of Humble Oil. (Most of Alyeska's employees were "borrowed" from the parent oil companies.) One of the first things Patton did as head of Alyeska was to make a speech before the Anchorage Chamber of Commerce in mid-September, at which he announced that the land claims had to be settled before there could be a pipeline. It was a remarkable change of attitude on the part of the oil companies and was the result of Judge Hart's injunction against a construction permit in the Stevens Village case and a summer of hard work by BP's Hugh Gallagher.

Gallagher and BP talked the American owner companies into hiring a full-time lobbyist in Washington. All along, Gallagher had in mind for the job another former Bartlett aide, a Washington lawyer with wide connections on Capitol Hill, William C. Foster. Although the job had really been created for Foster, he did not get it without a struggle since the American oil companies had their own ideas about who should represent them and many of the men they suggested were better qualified from the industry's point of view. Foster's unique qualifications included his knowledge of Alaska and his personal acquaintance with many of the people with whom he would have to work, including representatives of the state and the Natives. In the end, Gallagher prevailed and Foster was hired to represent Alyeska's interest in seeing the claims settled. In fact, Foster interpreted his mission very broadly and became involved in many aspects of the bill. His involvement was such that at several points Alaska's new Attorney General, John Havelock, relied upon Foster to translate the Native rhetoric for him so the state could understand exactly how much the Native leaders could compromise without losing their own following. But Foster did not enter the picture until after Congress adjourned for 1970.

September was a period of frantic but not particularly productive activity on the part of the oil companies. Frank K. Rickwood, president of BP Alaska, sent letters to every member of the House Interior Committee, urging that the claims "be speedily resolved in a manner that is fair and equitable to the Natives, to the state of Alaska and to the U.S." And several other companies let Chairman

Aspinall know that they wanted a settlement. Representative Aspinall, whose relations with the industry have always been cordial, got the message. In a private meeting with Native representatives in mid-September, the Chairman insisted that the Committee would report out a bill that month.

Back in Alaska, the Chambers of Commerce took Patton's lead and decided that any bill was better than no bill at all. On the evening of September 13, the top men in the Anchorage Chamber met with all the politicians who happened to be in town and humbly asked what they could do to help. Senator Stevens, who turned out to be the only member of the Congressional delegation there, was gloomy. There was not much chance that the House could come up with its own bill, he said. Better get them to pass the Senate bill or some version of it and objectionable parts could be changed in the final conference on the bill.

Then the Chamber group met with the Natives. Some Native leaders clearly thought the Chambers' humility had come a little late. "There's no way that just any old bill is going to get past the Congress just to satisfy people who are afraid their pocketbooks are going to get hurt," said one. But Don Wright, who became president of the AFN a few months later, was more conciliatory. "There's going to be a compromise," he said. "That's what America is all about." The result was the formation of an ad hoc lobbying group, Alaskans United.

Governor Miller, who was busy running for reelection, still sought to prevent the state from having to pay any part of the settlement, but indicated that he might accept a compromise in the interest of getting the North Slope oil to market. He planned to be in Washington when the House Indian Affairs Subcommittee had its first mark-up session on September 17, having won the Republican gubernatorial primary. So did Representative Pollock, having lost to Miller in the same primary. Senator Stevens promised to talk to President Nixon about the settlement and Senator Gravel promised to talk to Representative Haley.

In the interim, Representative Aspinall, who was really running Representative Haley's Indian Affairs Subcommittee, had not been idle. He set committee counsel Lewis Sigler quietly to work drafting a completely new bill. This draft bill would give the Natives $1 billion from a variety of sources and subsistence use of 40 million or more acres of land. In addition, the villagers would get title to the land immediately around their homes.

However, even with the draft bill before them, the subcommittee got nowhere. For one thing, it was late September and almost every member except Representative Pollock was up for reelection in six weeks. The panel was up against a deadline—the day, still unspecified, when Chairman Aspinall would leave for Colorado to campaign and all work would stop in the Committee. For four days, the subcommittee considered and rejected numerous plans, came to several tentative agreements, and got nothing down on paper. On September 22, it adjourned, ostensibly for the year, with Representative Haley promising vaguely to appoint a special task force to consider the claims further. Representative Pollock said angrily, "They're just playing games."

On the following day, the House Interior Committee met to wind up its business for the year. The only mention of the claims was in passing. "The only thing we haven't had is the funeral," said Representative John Saylor, laughing, to reporters before the session began. Chairman Aspinall was furious because a story had appeared in the *Washington Post* which said the Indian Affairs Subcommittee had agreed upon 40 million acres of land and a selection process which allowed the Natives to choose some land, the state some, the Natives some, the state some, and so on until everyone was taken care of. Representative Aspinall was convinced that the source of the *Post* story was Representative Pollock's legislative assistant, John Katz, who had sat in on most of the subcommittee's deliberations at Pollock's request. The Chairman was determined to punish Representative Pollock for this, although Katz actually was not the source of the story at all.

Officially the claims never came up that day. However, the Committee did not finish its business and agreed to meet the following morning. On September 24, the Committee wound up a few bits of pending business, chiefly a bill to allow leasing of federal geothermal resources which one member, Representative Craig Hosmer, R-Calif., had been trying to get passed for several years. Representative Saylor was in his usual form. At one point, the Pennsylvania Congressman was trying to amend one of Representative Hosmer's own amendments to the legislation, when Chairman Aspinall looked up at the clock which hung above his own portrait and remarked that time was running out for Representative Hosmer's bill. "You want a bill?" asked Representative Saylor arrogantly. He leaned way back in his big chair. The Californian looked mildly over the rims of his spectacles. "Since you're blackmailing me," he said, "I'll accept your amendment."

Later, Representative Saylor proposed that the Committee not meet again in 1970 except on matters of "great urgency." At this point, Representative Edmondson interjected, "We have pending a matter of extreme urgency in the case of the Alaska Native land claims bill." The Congressman from Oklahoma who had taken the *Anchorage Daily Times* to task for racism the year before, added, "I think there should be some discussion of what's going to be done." Representative Pollock, who had tried in vain to get permission to speak, looked over at his colleague gratefully. "From the standpoint of the state, it is a matter of the highest urgency," said Representative Edmondson.

Chairman Aspinall, who had announced earlier that he was leaving the next day for Colorado, entered the discussion. "The chair wants it known that he's going home," said Representative Aspinall. "The bill is before the subcommittee on Indian Affairs. If the gentleman from Florida, the Chairman of the subcommittee, wishes to hold more hearings, the chair will abide by his wishes. But the chair is in no position to force the gentleman from Florida to hold hearings." Representative Haley, perhaps anticipating a fracas, had left the room almost an hour earlier.

Representative Pollock tried again to be recognized. The Chairman ignored him and went on talking. First, he attacked the lobbyists who had been working on the bill. He would not be driven into action by state leaders, attorneys for the Natives, or Native leaders. Conspicuously, he did not mention oil companies. Next, he attacked the unnamed source of the *Post* story. "People have a tendency to foul their own nests," Representative Aspinall said with an ominous look at the Alaska Congressman. Then, he attacked the Natives for disagreeing among themselves about how the claims should be settled.

When Representative Pollock was finally allowed to speak, he pointed out that the claims had been passed over time and again for other legislation, that the AFN had presented a unified position on the claims, and that the land freeze was to end on December 31. He also raised the question of what would happen when it ended. The freeze could be extended, Representative Aspinall replied calmly. The Alaskan begged for time. The subcommittee's differences could be resolved in a few hours, he said.

"Don't point the finger about working," Representative Aspinall told him. "The gentleman was home to campaign. I'm going home to campaign."

The Alaskan made several more half-hearted tries; then Representative Edmondson took over. Would the Chairman be willing to write the President or at least the Secretary of the Interior to see if there had been any change in the administration's position on how much land the Natives should get? At this point, Representative Saylor interrupted, accusing Representative Edmondson of "playing politics" by trying to get the President to say something that conflicted with what the Secretary had said earlier. When Representative Edmondson persisted, Representative Saylor looked up at the clock and saw that it was after twelve. A House session had started at noon. Under House rules, a Committee may not meet while the House is in session unless it has special permission from the presiding officer. It is a rule which the Interior Committee observes strictly.

"I call a point of order," shouted Representative Saylor. "The House is in session."

Then he got up and strode toward the door. To leave the room, he had to pass Representative Pollock's chair. The Alaskan half rose to meet him. "If you'd tried to help me anywhere along the line during the past four years, you'd have gotten my help," sneered Representative Saylor.

"You haven't been any leader as far as I'm concerned," replied Representative Pollock, leaping to his feet.

"Nuts to you," said Representative Saylor, "I'll be around next year and you won't." The two men glared at each other.

"John, John," interjected Representative Hosmer, who was standing near the pair. Representative Saylor strode out of the room.

Some Natives and one of their lawyers, Ed Weinberg, were watching. "What a way for it to end," remarked Weinberg.

That was the last of the claims bill in the Ninety-first Congress. Half-hearted attempts were made to resurrect the settlement but most were really designed to lay the groundwork for 1971.

In November, the Democrats swept into office in Alaska. Representative Pollock was replaced by a shrewd, intensely ambitious school administrator named Nick J. Begich. Governor William A. Egan, the state's first Governor, was reelected, bringing with him John Havelock as Attorney General. The Governor made Havelock his chief negotiator with the Natives and the two men set about to heal the rift between Native and non-Native in the state.

The Natives were having problems of their own. The militants had never been happy with the Senate settlement bill, and the land question led to the first major public split in the AFN's ranks since its founding in 1966. Charlie Edwardsen, now executive director of the Arctic Slope Native Association, announced at the annual convention of the National Congress of American Indians held in Anchorage that the North Slope Eskimos were withdrawing from the Alaska Federation of Natives to seek their own settlement. "We believe it no longer remains possible for the ASNA to sit in the AFN councils," said Edwardsen. "The AFN has lost sight of the fundamental principles upon which the entire settlement is premised. That is, this is a land settlement, not a federal welfare program or another piece of anti-poverty legislation."

Eben Hopson, a former ASNA executive director himself and still acting head of the AFN, was caught in the middle. He tried to stop Edwardsen from making his announcement or at least to postpone it. But he understood his feelings. "It just goes to show you that I no longer represent the ASNA," he remarked a little sadly. In addition to all 40 million acres, the ASNA wanted the land distributed according to the amount used and occupied historically by each Native group. Since the ASNA had made the largest claim, under their proposal they would receive more land than any other Native group, roughly 15 percent of a 40-million-acre settlement.

The convention ended with the ASNA officially withdrawn from the AFN's ranks and no decision made by the AFN board as to what to do about it. That was up to the new President, Don Wright, the son of an Athapascan Indian from Tanana and an Episcopal missionary nurse. A skilled politician, Wright wisely let the matter ride for a while.

In early December, the AFN board met to decide what to do. After three days of debate, the board finally voted to raise the land portion of their settlement proposal to 60 million acres, after rejecting 100 million as an impossible goal, in order to accommodate the North Slope Natives. The state was to be divided into twelve ethnic regions. Both the land and the money were to be distributed to these regions on the basis of land lost. Fifty percent of the revenues from oil and mineral development were to be retained by the region in which the development occurred. As a result, the ASNA returned to the fold, as they had apparently intended to do all along.

Their walkout had been a successful ploy. It had forced the AFN to reassess its position and ask for more land. "They never really broke away," Wright later said. "They just had a position they felt was never fully understood or considered by the AFN.

They felt the AFN had not done all it could to get a good bill through Congress. The ASNA felt their emphasis that this was a land claims settlement was not fully understood by everyone in the AFN. They made their point at the meeting and now we're heading in the right direction." [101] Wright himself had just returned from a long trip to Washington, where Congressional and administration officials had warned him that he had better bring his people together. Apparently unified once more, the AFN prepared for conferences with the new Governor and his Attorney General.

Governor Egan had not been idle either since his election. Almost as soon as the ballots were counted, he announced that he would go to Washington to see if anything could be done about the claims bill that year. Governor Egan was no fool. He knew when he left Alaska that the legislation didn't have a prayer in the Ninety-first Congress but he wanted to renew old acquaintances on the House Interior Committee and repair some of the damage which had occurred. However, Representative Aspinall was not in Washington. A widower at seventy-five, he had remarried the summer before and, once safely reelected, had taken his bride on a round-the-world wedding trip.

Governor Egan came to Washington anyway, accompanied by the new Alaskan Congressman, Begich. The Governor, a Valdez grocer before entering politics and a genuinely humble man, was perfectly suited to the role he had to play. He did what he was expected to do. Substantive matters were rarely discussed at these and later meetings Governor Egan had with the hierarchy of the House Interior Committee. Committee counsel Sigler usually handled substantive things. Instead the Governor and the Chairman talked about trivia. At one meeting, they discussed the Chairman's age and whether it showed. Another time the Governor asked when the Chairman planned to schedule a meeting. "In my quasi-judicial position as Chairman of the Committee," replied Representative Aspinall, "I can't give you that information."

Governor Egan's message during that first visit in the late fall of 1970 was that his administration was flexible about a settlement.

"Flexible" had been a key word during his recent campaign. By being "flexible," the Governor avoided ever saying exactly what he wanted as a settlement and so avoided committing himself too far in any one direction. His was a masterly performance and Representative Begich took his cue from it. Even when the Natives tried to get him to support the new AFN proposal, which called for 60 million acres, Representative Begich refused.

During Governor Egan's 1970 trip to Washington, Senator Gravel gave a party for him. He invited everyone in Washington with Alaskan connections, everyone who had an interest in the settlement of the claims or the construction of the pipeline, to a buffet at the Tantallon Country Club near his home. Tantallon is one of the many high-priced developments which have sprung up around Washington in recent years. It is on the Potomac River and the country club has a view of Mount Vernon on the opposite bank. After Senator Gravel decided to give the party, he looked around for someone to pick up the tab. It was short notice, but he was able to get several oil companies to foot part of the bill.

Short notice or not, everyone came, if only to see who else had been invited and to muse about why. It was a party that harkened back to the good old days of the statehood battle. There were Senators, Congressmen, their wives and aides, members of the Interior Committees' staffs, important people from the Interior Department, Natives, oil company representatives, former Senator Gruening and his wife, and the late Senator Bartlett's daughter. In fact, the only notable absences were those of Interior Secretary Hickel, who was making a speech in New York that night, and Chairman Aspinall, who was on his honeymoon. The party was a success. The good feeling generated there lasted a long time. Remarked a lobbyist who was present, "That dinner Mike gave was a good thing."

Back in Alaska, Governor Egan met with Native leaders to smooth the feathers his predecessor Miller had ruffled. On December 20, the Natives and the Governor met all day and emerged in the evening with smiles. Likening the land claims fight to the statehood struggle, the Governor said that both sides now understood one another. "This was history," remarked Hopson. Native leaders Wright and Borbridge talked about real estate transactions and economic development. They said this was what it was all about and that the state finally understood this. In truth, neither side had given in on any specific points. But each came away feeling it had won. The meeting was really a victory for Governor Egan's policy of "flexibility."

In the meantime, President Nixon had fired Secretary Hickel. Despite the Secretary's departure, work continued in the Interior Department on a draft environmental impact statement on the pipeline. The idea had been to produce this draft, submit it to other government agencies and the public for comments, hold public hearings on it, and then combine the draft and the comments into a final statement, which, it was hoped, would satisfy the courts. When Secretary Hickel was fired, his subordinates immediately pushed some of his favorite projects through the department's red tape but, much as they might have liked to do so, there was no way to get the pipeline approved before Hickel left town. The draft impact statement could not be finished before Christmas and Secretary Hickel was fired the day before Thanksgiving. Although somewhat in limbo after the Alaskan's departure, a task force headed by Jack O. Horton, a young assistant who had been brought into the department by Train but remained after Train moved over to the Council on Environmental Quality, continued its work.

At the beginning of 1971, the Alaska Natives prepared to approach a new Congress with new demands and new allies, the oil companies. The oil companies put the heavy equipment they had strung out across the Alaskan wilderness into mothballs for another winter and deliberated on how best to approach the new Secretary of the Interior, Rogers C. B. Morton.

8

THE MILLION-ACRE KISSES

"It is long past the time that the Indian policies of the federal government began to recognize and build upon the capacities and insights of the Indian people. . . . The time has come to break decisively with the past and to create conditions for a new era in which the Indian future is determined by Indian acts and Indian decisions."

President Nixon, July 8, 1970
message to Congress on Indian
Self-Determination

On February 16, 1971, the Commerce Department auditorium was packed to overflowing. Spectators leaned forward in their sharply banked seats. Politicians, lobbyists, and reporters mingled in the well and the halls outside. Jack O. Horton, then an assistant to the Secretary of the Interior, banged his gavel and called for order. The first major public National Environmental Policy Act hearings had begun. Their subject: the controversial Trans Alaska Pipeline.

A 196-page draft environmental impact statement, required by Section 102 (c) (2) of NEPA, had been released in January. In addition to circulating this document to interested government agencies as the law required, the Interior Department scheduled hearings on it. But the Department underestimated public interest in the project. The list of witnesses grew and grew. At the last minute, the hearing had to be moved from the cramped Interior Department auditorium to a larger one at the Commerce Department.

The hearings continued day and night for three days. More than a hundred witnesses appeared: politicians, government officials, oilmen, petroleum consultants, Alaska Natives, Alaska businessmen, professors, housewives, students, professional conservationists, and even a few inventors, one of whom brought along a model of an ingenious device to insure the pipeline's integrity. Even more spectacular hearings were held in Alaska a week later—more all-night sessions, more witnesses. By the end of March, the Interior Department, which was holding the hearings, had amassed almost 3,000 pages of transcript, another 3,000 pages of appendices, and some 1,000 pages of letters and comments. The Department was very proud of the bulk of this record, as though its weight alone was proof that the Department had done its job well.

But the content was mostly critical of the pipeline. Some criticism was philosophical, some was technical. And there was a theme which recurred throughout: the Department had first decided to build the pipeline and then set about to justify the decision. The Interior Department was merely playing games with NEPA.

There were some exciting moments during the Washington hearings. Governor Egan got up the first day and announced that the state of Alaska would be broke by 1976 if the pipeline was not built. An articulate member of the Canadian Parliament from Vancouver, David Anderson, speaking for himself and not for the government of Canada, called attention to the dangers of heavy tanker traffic to Canadian fisheries and the possibility of oil spills in the northern Pacific. Charlie Edwardsen from Barrow and Chief Richard Frank from New Minto held a press conference, at which Edwardsen taunted the oil companies, "My people would like no oil development, and if there is oil development, let the pigs pay their rent." Former Interior Secretary Stewart Udall warned that Alaska "will select all the valleys and give the federal government the mountains and throw the whole thing open to oil." Representative Les Aspin, D-Wis., released barely legible copies of an internal Interior Department memorandum by a BLM employee named Harold Jorgenson, which criticized the Department for ignoring possible environmental damage out of deference to the oil industry. Only the day before, Jack Anderson had published excerpts from Jorgenson's memo in his daily column.

Jorgenson's criticism was echoed by James Moorman, then chief attorney for the conservation groups which had obtained an injunction against the project the year before. Moorman objected to the draft's "inappropriate spirit of advocacy for the pipeline." Then he went on to list areas in which he, his clients, and the scientists they had consulted, had found the statement lacking. They were numerous. Moorman wondered why the Department had made no mention of what would happen to the oil after it left Valdez on the southern coast of Alaska. Clearly, the gap between Valdez and the contiguous United States was the Achilles heel of the whole oil transportation system. He also wondered why there was no discussion of alternative routes, particularly of the possibility of a pipeline up the Mackenzie River Valley through Canada and eventually to the American midwest. He wanted to know why the Alyeska Pipeline Service Company had provided no specific designs. And he asked about the width of the proposed right-of-way. True, the company was now asking for the legal fifty-four foot right-of-way, but they were also requesting numerous special use permits that would increase its size markedly.

Oilmen and Interior Department officials had expected criticism. But what came as a surprise to them was the number of private citizens who took time off to come to Washington or Anchorage to testify and the reasoned, well-informed nature of their comments. One Alyeska official said later, "I was both depressed and impressed." On the second night of the hearings, while they were in progress, Senator Stevens held an Alaskan party in his suburban Maryland home. The atmosphere was quite different from that at Senator Gravel's party a few months earlier. The men who had been attending the hearings—Governor Egan and his party of state officials and Alyeska president Patton and his coterie of vice-presidents and consultants—were feeling the full force of public opinion against them. "I think," said one oilman, "it all spells D-E-L-A-Y."

It was not as if the state and the oil companies had not tried to gain public sympathy. After its creation in the late summer of 1970, Alyeska had embarked on a costly program to reshape the public image of the pipeline. It included a series of handsome, slick advertisements in most major newspapers and magazines across the country in late 1970 and early 1971. The first showed an Eskimo in a parka holding up a measuring glass full of a dark liquid, presumably crude oil, with a drilling rig in the background. "DO WE REALLY NEED IT?" asked the headline. America uses 14 million barrels of oil a day, the ad read, but produces only 10 million. The gap will get wider and the country will be forced to import more and more oil from "politically unstable areas," it predicted. "North Slope oil will also bring economic blessings to the state of Alaska—it'll mean jobs and opportunity for all Alaskans—particularly Native Alaskans. In short, the oil is a boon." The ad went on to discuss the formation of the new pipeline company and the cost of the project to date, "the most expensive single project ever undertaken by private industry." "We accept the fact that some 205 million Americans will be looking over our shoulders to make sure we do the job right. We know environmental problems must be faced—and solved—before the project can proceed. It will take a great deal of care and a great deal of money, but we know we can build the pipeline without significant damage to the land or to the wildlife. . . . Now we think it is time to move ahead."

A second advertisement showed a caribou suckling her calf with the legend: "WILL A PIPELINE RUIN THEIR ARCTIC?" This one was based upon the companies' original premise, that most of the pipeline could be buried safely. The copy implied that the route had been selected because it passed through areas of ice-free permafrost. It then turned to the problem of the caribou: would the pipeline present an insurmountable barrier to their natural migrations across the North Slope? Still assuming that most of the pipe could be put underground, the ad assured readers that no above-ground section would be more than forty miles long, a distance a caribou can easily go in a day, according to Alyeska. The ad concluded:

> What we have learned about the Arctic leads us to believe that there is nothing inherently dangerous to the environment providing the line is designed, built and operated in a manner that is considerate of and responsible to the environment. In truth, what's good for the environment is also very good for the safety and security of the pipeline. On this you have our pledge: the environmental disturbances will be avoided where possible, held to a minimum where unavoidable, and restored to the fullest practicable extent. And we can assure you that the pipeline will be the most carefully engineered and constructed crude oil pipeline in the world.

In April 1971, the American Petroleum Institute (API) launched a $4 million advertising campaign around the theme, "A country that runs on oil can't afford to run short." The API used nighttime television, plus spots on the Today Show and space in more than 180 newspapers and magazines, to emphasize not pollution control programs but the need to develop Alaskan reserves to lessen United States dependence on foreign oil. This "educational" program began after API took a poll which showed that the public had little understanding of, or sympathy with, the industry. One spot, filmed in Grinnell, Iowa, showed a housewife making coffee, a doctor with a hypodermic needle, and a grandmother at her sewing machine. The idea was to show people that petroleum had all sorts of uses. But there were no automobiles in the ad, although nearly half the petroleum used in the United

States runs them. Furthermore, two million barrels a day, the projected throughput of the Alaskan pipeline at capacity, is but a drop in the bucket of United States petroleum needs. Based on projections made by the Interior Department for its final environmental impact statement for the Alaskan project, the United States will need about 22 million barrels a day by 1980.

Standard Oil of New Jersey prepared a series of television advertisements, which appeared on nightly newscasts and public affairs shows like NBC's "Meet the Press." The commercials contained favorable references to the pipeline or discussions of Esso's experiences in the Canadian Arctic, where the company had been drilling through its majority-owned subsidiary, Imperial Oil. The Federal Communications Commission later ruled that these ads violated the fairness doctrine because they showed only one side of an important public issue.

Governor Egan's administration was also taken aback at public reaction to the proposed pipeline, though some Alaskans felt that this was the usual sort of misguided interest in Alaska exhibited by people from Outside. Under Governor Miller, the state had made a feeble attempt to win friends for itself, but the so-called Alaska Plan never really got off the ground. The Alaska Plan consisted chiefly of a booklet issued in 1970, in which the Governor promised that a century later wolves would still "roam open desolate tundra in abundance, thriving off the great herds of delicate caribou as nature provided." Later, the state produced a traveling exhibit of pictures and other display materials to be shown at various environmental conferences across the country. There was also the infamous "fake fish film," a documentary entitled *New for Tomorrow*, which featured a sockeye salmon swimming in a North Slope stream even though sockeye do not live in the Arctic. This film, reportedly paid for by several oil companies and produced by the state, purported to show that oil development would not endanger the wildlife on the North Slope.

During Governor Egan's administration, concern for Alaska's image eventually led to an operation called Speak Up Alaska! In

April, the Governor named a special task force to counteract unfavorable Outside publicity about the pipeline and the state, but it bogged down over financing. In May, Bob Reeve, "the Glacier Pilot" and founder of Reeve Aleutian Airways, formed his own committee and designated May 29 as "Speak Up Alaska Day." Reeve sought 10,000 Alaskan participants. All you had to do was fill out a blank, listing states in which you knew people who would be willing to publicize Alaska. The program had the endorsement of several major newspapers and Senator Stevens, but it came to nothing in the end.

As oilmen and state officials moodily munched lasagna and salad at Senator Stevens' house that chilly February evening, the Natives present were buoyant. By now, they had the oil companies over a barrel: no settlement, no pipeline. The public, even the conservationists, seemed to be on their side. And there were indications that the Nixon administration might agree to the sort of settlement the AFN was seeking in order to exemplify the "Indian self-determination" policy proclaimed by the President the previous summer. That night, when Charlie Edwardsen declined a glass of wine offered by a waiter and instead took one from Ed Patton's hand, sipped it, and then returned it, it was with the deliberation and delight of a man who knows that, at last, his hour has come.

As far as the Natives were concerned, the emphasis in the claims controversy had shifted from whether the state should participate in the claims settlement at all to how much land the Natives should get and how that land should be chosen. Having finally resigned themselves to the state's paying the Natives a two percent mineral lease royalty, both the state government and the Alaskan business establishment turned their attention to how the state could get the best land. Land selection was a far more sophisticated issue than the royalty.

The state's strategy throughout the Congressional maneuvers of 1971 was based on the following: the state wanted any land grant tied as closely as possible to the sites of the present villages.

The state did not want the Natives to be able to select land anywhere, strictly for its economic potential, as the AFN proposed. Limiting the Natives to land directly around their villages would mean giving them some land, but it would preserve for the state large areas like Prudhoe Bay which were far from any village. Many villages were located on land that contained no minerals or oil.

The state had another reason for wanting to tie the land to the villages. Such a settlement would be a village settlement. The land ownership would be centered in 180 or more Native villages, not in regional corporations. This meant that Native political power would be diffuse. On the other hand, twelve regional corporations, as the AFN now proposed, molded after the existing Native power structure and having land and mineral resources at their disposal, would be formidable economic and political forces in a state with a population of scarcely more than 300,000 persons. But a couple of hundred Native villages would have little more political clout than they had at present. Alaskan politicians were fearful of upsetting the existing power structure in any way that would favor the Natives. They talked about dreading "a state within a state" but what they feared was a disruption of the political status quo.

The state also wanted the settlement to end the land freeze immediately, so it could finish choosing its 102 million acres and so that its non-Native citizens could prove up existing claims, get their leases, and go on with the business of developing Alaska.

But there were tricky questions, like who would get to choose first after the settlement, the state or the Natives, and what would happen if the federal government wanted to keep some of the public domain for itself. Governor Egan and Representative Begich saw huge Naval Petroleum Reserve #4 as an answer to their prayers. Here were 23 million acres of potential oil fields that the state could not get its hands on. Since there was no chance for the state to have Pet Four, why not give it to the Natives? Furthermore, the Natives on the North Slope wanted Pet Four. It was an ingenious idea, for it allowed the Egan administration to support a generous land grant to the Natives while giving up very little valuable land that could reasonably be expected to become Alaska's. The problem was that it was unlikely that Congress would ever agree to it. At one point, Native leaders put out tentative feelers to Representative Edward Hébert, D.-La., Chairman of the House Armed Services Committee, which had jurisdiction over Pet Four. Representative Hébert told them he was sure that Chairman Aspinall would never allow his Interior Committee to approve such a thing, and let it be known that if he did, Representative Hébert would move to have the claims bill sent to his Committee, thus killing it. The threat was enough. The plan to use Pet Four in the settlement came to nothing.

Since Governor Egan's election, the state had acquired at least the façade of a unified position. On the other hand, the Natives were far from unified. Despite the return of the ASNA to the AFN fold, the argument over the distribution of land and money continued. The ASNA wanted the money distributed on the basis of land claimed. The Tlingit-Haida Central Council wanted it distributed on a per capita basis. The Tlingit-Haidas finally won out, largely because the northern Eskimos cared more about the land than the money.

The Natives were also having problems with their lawyers. They blamed Ramsey Clark for the poor settlement the Senate approved in 1970 and accused him of selling them out because he was too busy with other clients and because he had political ambitions of his own. By this time, former Justice Goldberg was completely out of the picture, having withdrawn to make his unsuccessful run for Governor of New York. So was Clark. Their firm was henceforth represented by Kenneth C. Bass III. Ed Weinberg remained on the case. The Natives' renewed contracts with their Washington lawyers specified that a member of each firm be present at all AFN meetings and in constant contact with the Natives.

In the beginning of 1971, Congress had several possible settlements to consider. One was the Senate bill of the year before, legislation the Natives swore they would not accept. In a January 1971 memorandum to the Interior Department, AFN consultant Adrian Parmeter called it "an unmitigated disaster" and warned that the Natives were looking to the administration for leadership. Wrote Parmeter, "Without 40 million acres of land [with full title] the President's policy of self-determination of American Indians is a hollow slogan indeed." [102]

Early in 1971, Senator Jackson reintroduced the 1970 Senate bill with the controversial Section 17 on mineral leasing and the hated Five-Year Freeze. He asked the two Alaskan Senators to cosponsor it as a matrix from which the Senate Interior Committee could work, and they did. Their cosponsorship was widely misinterpreted by non-Native Alaskans as endorsement of the bill's contents. Letters and telegrams poured into the Senators' offices, many of them inspired by an Alaska Miners Association advertisement which listed the members of the Committee by name and offered form letters of opposition to the Jackson bill to anyone who could not compose his own. Two miners wired Senator Stevens: "Your name on Senate bill 35 which contains a five year land freeze and competitive mineral leasing is shocking. These two provisions will set the state back 25 years and your political future about the same."

Besides the Senate bill, a second proposal before Congress was the bill Representative Aspinall had had his staff draft at the end of 1970. In early February, Representative Aspinall introduced it in a characteristic way. He told no one what he was doing, not even Alaska's new Congressman, Nick Begich, a member of his Committee. (The oil industry, however, knew about the bill before it was introduced.) What Representative Aspinall proposed to do was to give each village full title to the land it presently occupied plus adjacent land equal to three times its present size, from one to three million acres in all. Then the Secretary of the Interior would give subsistence use permits to the Natives, allowing them to hunt and fish on at least 40 million acres. Finally, the Natives would get $1 billion, half of which would come out of state mineral revenues.

The Aspinall bill was not at all what the Natives wanted. It would have meant even less land than they would have gotten under the Senate bill. And they would have been dependent upon the Secretary of the Interior for use of the land.

Therefore, the Native leadership was alarmed when the new Secretary of the Interior, Rogers C. B. Morton, appeared before the Senate Interior Committee two weeks after Representative Aspinall introduced his bill and indicated that his department was drafting a similar bill. Secretary Morton had been a member of the House Interior Committee until he was appointed to the Cabinet and was a friend of Chairman Aspinall. Don Wright, who had met with the new Secretary shortly after he took office, wrote him anxiously, "I and my fellow Alaska Natives were frankly shocked at your testimony last week. . . . In our opinion, the legislation which you indicated you will submit is a serious retreat from last year's administration position. . . . This bill which you outlined has obviously been prepared hastily and without a full understanding of our rights." [103] Later, Wright wrote President Nixon in the same vein.

Meanwhile, the AFN had persuaded Senator Harris, who had been helpful to the Natives during the 1970 Senate debate, to introduce a new AFN bill. Senator Harris's bill, which was introduced in the House by Representative Lloyd Meeds, D-Wash., gave the Natives 60 million acres, to be divided among twelve regional organizations. The regional corporations, in turn, would give each village the land on which it was located plus additional acreage for expansion. But most of the land would remain in the hands of the regional corporations. The AFN bill spelled out the way in which the land would be allocated among the regional corporations. Fifteen percent of the land, about nine million acres, would go to the North Slope people; eight percent, or nearly five million acres, to the Indians in the southeast; five percent, or three million, to the Aleuts; and so on throughout the state. This distribution plan reflected Edwardsen's criticism that the land ought to go to those who had historically used and occupied it.

It had been Senator Jackson's intention to hold one brief hearing at which each interested party would have a chance to update his case. He wanted to hear from Don Wright, Governor Egan, and Secretary Morton. But when Secretary Morton came up to Capitol Hill on February 18 and indicated that he was thinking along the same lines as Representative Aspinall, he had only been in office a short time. He begged for more time on the claims. At the same time, Governor Egan and the Natives were unwilling to commit themselves before hearing what he had to say because they knew that steps were being taken to turn the administration around on the land claims.

The key was the President's July 8, 1970 message on Indian self-determination. The National Congress of American Indians (NCAI) was holding a conference on Indian self-determination in Kansas City in early March. Vice-President Spiro T. Agnew would be attending this meeting as nominal head of a federal organization called the National Council on Indian Opportunity. The Natives hoped to corner the Vice-President at the meeting and win his support and that of his council. Before the meeting, Wright wrote him: "We are outraged over Secretary Morton's recent statements. Unless there is visible proof that this administration means what it says about dealing fairly with the Natives, I fear that our people might not continue to pursue the peaceful path of negotiation." (He was probably thinking of Edwardsen, who had recently threatened to a Washington oil lobbyist that he would blow up the proposed pipeline.) Wright did not stop with letters. In late February, he met with both C. D. Ward, one of the Vice-President's domestic advisers, and Robert Robertson, executive director of Agnew's Council on Indian Opportunity. On March 3, as the Vice-President was en route to Hawaii to make a speech on revenue sharing, someone on Air Force Two placed a call to the White House to learn the details of the land claims legislation the administration was considering. A few days later, Vice-President Agnew was in Kansas City for the National Congress of American Indians. In an early meeting with eight Indian members of the Council, the Alaskans' land claims were the chief topic of conversation.

One of the members of the Council was an Eskimo woman from Fairbanks, Laura Bergt. Mrs. Bergt, who was born and raised in Kotzebue, was the wife of a white officer of Interior Airways, the airline which had handled most of the 1968–69 airlift of equipment to the North Slope. Laura Bergt was an Eskimo who had made it in both worlds. An attractive, poised woman, she was frequently found at Alaskan trade fairs, modeling dresses made of musk ox quiviut, a Native cottage craft which the University of Alaska was encouraging. When the Vice-President, then a candidate, visited Alaska during the 1968 campaign, he was greeted at the Anchorage airport by some Eskimo blanket tossers. Blanket tossing is an old Eskimo ritual, now usually displayed for tourists visiting Barrow and Kotzebue. It was Laura Bergt who was tossed in a blanket that day for Vice-President Agnew's entertainment. At the NCAI meeting in Kansas City, Mrs. Bergt renewed her acquaintance with the Vice-President by teaching him how to kiss in Eskimo fashion, that is, by rubbing noses. Borbridge, watching her, remarked that each one of those kisses was worth a million acres. And they were, for Mrs. Bergt was deadly serious. At her urging and that of other Natives present, the Vice-President set up a meeting with Native leaders and Interior Department officials in his Executive Office Building office near the White House on March 12. At the same time, the Congress of American Indians supported the AFN position and sent a resolution to this effect back to Washington.

Charlie Edwardsen was also active in Kansas City. He followed Vice-President Agnew, the convention's lead-off speaker, with an impassioned plea, "The only power that we have to exercise is the power of saying no!" Edwardsen attended one of the meetings between Agnew's people and the AFN board members (Edwardsen was not a member of the AFN board) and, fearing that the other Natives would compromise with the administration, brought his fist down on the table with a bang. "There will be no compromise," he shouted. "No. No." Then he walked out of the meeting, leaving a stunned collection of administrative assistants and Natives to pick up the pieces.

Meanwhile, as the Natives prepared to woo the administration, sometime before the NCAI meeting Senator Stevens, the only member of the Alaskan Congressional delegation who was of the President's party, had called the President's chief domestic adviser, John Ehrlichman, to say that the claims settlement was the first real test of the President's self-determination policy. The Senator had been struck by phrases in Parmeter's January memorandum for Interior Department officials. As Senator Stevens remembers it, he also told Ehrlichman at that time that he had thought of a way the Natives could be given 40 million acres without damaging the state's or the federal government's interests in Alaska. What Senator Stevens sketched briefly to Ehrlichman was a settlement in which the land grant was tied closely to the villages. Later, the Senator had an elaborate series of maps and transparent overlays made up by the state Division of Lands which showed the land selected by the state prior to the freeze, the land set aside by the federal government in specific reservations, and the locations of most of the Native villages. Ehrlichman listened to Senator Stevens, generally a loyal supporter of the President, and then asked Leonard Garment, who was in charge of minority issues, to assign a member of his staff to work with Senator Stevens and the Interior Department on a claims bill that both the Senator and the AFN could support.

Late on Friday afternoon, March 12, Vice-President Agnew, Secretary Morton, Assistant Secretary of the Interior for Public Land Management Harrison Loesch, Bureau of Land Management chief Boyd L. Rasmussen, assistant director of the Office of Management and the Budget Donald B. Rice, and the Department's legislative counsel, Frank Bracken, met with Wright, Al Ketzler of the Tanana Chiefs, Laura Bergt, and Ray Christiansen, an Eskimo State Senator from Nome. A variety of aides also attended the meeting, including Barbara D. Kilberg, a Domestic Council aide to Ehrlichman, Bradley H. Patterson, Jr. of Garment's staff, Vice-President Agnew's domestic assistant, Ward, and Robertson from the Council on Indian Opportunity. Charlie Edwardsen was not admitted to this parley. Wright argued eloquently for his people, presenting their case from the days of Russian occupation to the present. Everyone at the meeting was greatly moved. No substantive promises were made that afternoon, but Interior Department officials Loesch and Bracken indicated that the Natives would be allowed to participate in the drafting of the administration bill. Everyone present knew that a turning point had been reached in the Natives' long journey.

The following Monday, the AFN board met in Washington to discuss a Congressional lobbying effort. What evolved at this meeting was a lobbying pattern the Natives used effectively throughout 1971. They divided into teams of three or four persons, mostly residents of the same village or region, and made personal visits to as many Senators and Congressmen as would see them.

The big stumbling block was Chairman Aspinall. Several AFN consultants and attorneys who knew the Chairman offered advice to the board. "He hates long meetings," said one of them. "Avoid any mention of the Senate," warned another. "The House is the important body." "Make it plain that his bill is not adequate to meet the needs of the Natives," volunteered a third. Then they discussed who should go to see Aspinall. "No more than four," said an attorney, "all Natives." He suggested Laura Bergt. Wright had already thought about that. He recommended that he, State Senator Christiansen, Mrs. Bergt, Charlie Edwardsen, and Frank Degnan from Unalakleet go to see the Chairman.

This left the 534 other members of Congress. One of the questions which had to be resolved was the degree to which each regional team should argue for its own particular interests, a point of great concern to the North Slope Natives, who were represented at the board meeting by Edwardsen and Joe Upicksoun. "We are going to work discreetly for amendments," said Upicksoun, who had succeeded Edwardsen as executive director of the Arctic Slope Natives Association.

But other members of the board, while accepting the premise that each region had a right to its own land, were worried that the AFN would not be presenting a sufficiently united front to Congress. (One reason for having each team represent a region apparently was so that none of the interregional arguments should surface during the lobbying process.) But soon the discussion gave way to exhortations by Edwardsen and Upicksoun that the Natives walk confidently into Congressmen's offices and demand their land.

"You fellows are the land owners," said Upicksoun. "This is your land, you have been there forever and no Caucasian. Hell, no. They said, man you are savages. We have a better thing for you. Bull shit."

"The only reason we are here is because we own land," continued Edwardsen. "Up until this point, we have been circumcised by people who collaborated with the state and federal bureaucracy and the oil companies. And the only reason and foremost cause all of you are here is your aboriginal title and, until extinguishment, we have to shoot for the moon."

Said Upicksoun, "Your title is going to be extinguished forever. That's it brother, no more. I said this at the AFN convention. You are landlords and you god damn better look like landlords."

"The state of Alaska is your worst enemy," said Edwardsen. "We may be a part of the state by happenstance. The Tlingits were the losers. They only got a lousy seven and a half million dollars. Fight for the land in these lobby efforts. The heat is on, baby. Don't be nice to Congress and the Senators. And put your backbone up. No one is going to compromise no more to the state."

Upicksoun concluded the pep talk with a suggestion as to how lobbyists could use the pipeline in their arguments. "There will be no pipeline until this question of land is settled," he said, "because all the pipe setting in Alaska, whose land is it setting on? Whose land are they laying the pipeline on? It is our land, fellows, until

Congress decides to extinguish our title to it." Then the board members divided themselves into five teams and assigned each team eight or ten Senators to visit.

The next step in the Natives' Congressional strategy was to pressure the Alaskan delegation into supporting the AFN bill. All three members were invited to attend the meeting at the Capitol Hill Hotel. But only Representative Begich came and he refused to be pinned down. Taking an extreme position and being belligerent about it wasn't the way to handle Representative Aspinall, the Congressman argued. "I'm a freshman," he told the Natives, "a new man. These men run the ball game and you make your points at the right time. The game is a matter of timing . . . and the results are what you get in the end." At this point, Edwardsen brought his fist down so hard on the table in front of him that it collapsed, dumping the glasses upon it onto the floor. Everyone was stunned, but Wright asked the Congressman to finish speaking. "I've got to weigh things carefully," continued Representative Begich. "I want to get the best bill out. But I don't want to be a martyr."

The AFN was not interested in compromise, Wright told Representative Begich. When the Congressman's legislative assistant, who had accompanied him to the meeting, tried to explain that in Congressional negotiations it was important not to get locked into an extreme position, the Natives would have none of it. They remembered that the year before they had left the details to their Congressional delegation and the Senate had not given them what they wanted. Then, at Edwardsen's insistence, the board voted that the whole Alaskan Congressional delegation be urged to support the AFN bill, and Representative Begich left saying, "Reserve judgment until I perform. Condemn me if I am wrong, but give me a chance to produce. . . ."

Representative Begich left the meeting angry. He had always kept the Natives at a distance but he felt even more removed than ever. Part of the Congressman's reserve about the Natives was tactical; part was due to fear that Wright would run against him for

Congress. Wright had once opposed Representative Begich in a primary election and the Congressman was convinced that he would do it again. But he was also bewildered by what had happened at the meeting.

During the next few days, teams of Natives toured the Senate and House office buildings, rounding up support for their cause. The teams saw most of the Senators, or at least members of their staffs, and about 180 Congressmen. Then most of the board members had to go home. It was left to Native leaders like Wright and Edwardsen, who remained in Washington all year, and a group of volunteers called Alaskans on the Potomac to make contact with the rest of the Congressmen until the last few days before the bill came up for a vote in the House, when Native leaders again visited Capitol Hill in force.

While the Natives were discussing lobbying tactics, the Interior Department notified Senator Jackson that Secretary Morton would not be able to tell his Committee the administration position on the claims on March 16 because of the White House's intervention. The Natives did not learn this until Tuesday morning, shortly before the Senate Interior Committee hearing was to begin. It was an odd hearing. Senator Harris was expected to explain the AFN bill he had introduced but had gone instead to civil rights leader Whitney Young's funeral. Secretary Morton did not come. The Native leaders did come, but refused to testify. In the end, Senator Jackson agreed to put off the hearing until the administration had sent up a bill.

The next three weeks were hectic ones for the Native leadership, their lawyers, the Interior Department officials responsible for drafting the claims bill, Senator Stevens, and Brad Patterson, the White House aide assigned to work on the bill. The Senator took his maps and overlays down to the Interior Department, where they were the subject of frequent consultations. The Native leaders were in and out of the Department and Senator Stevens' office.

The first problem was land. Sixty million acres seemed huge. Various other figures were batted around, but this was where

Senator Stevens' maps were important. With them, the Senator was able to prove that the Natives could be given 40 million acres without taking too much land that the state or the federal government wanted. In Senator Stevens' mind the key was that the land should be around the villages. The AFN continued to hold out for land anywhere in the state or what became known in the legislative parlance of the settlement as free floating selections. Eventually, the administration agreed to give the Natives 40 million acres of land, mostly contiguous to the villages.

The administration bill also included $1 billion—$500 million in federal appropriations and $500 million from a two percent royalty. The powerful Office of Management and the Budget had opposed a royalty, as had its predecessor, the Bureau of the Budget. However, the White House insisted upon it. It was the President's attitude that carried the day. Although he never met with Wright or any other Native leaders until the day the White House formally announced what was in the administration bill, President Nixon let it be known that he wanted the sentiments of his self-determination message to be embodied in the claims settlement. The head of the Domestic Council, John Ehrlichman, who spoke for the President, was the final arbiter in the drafting process.

One concept the Natives cherished but were unable to sell to the administration was that of regional corporate organization. Despite pleas to Ehrlichman through Senator Stevens, the Natives lost.

In form, the administration bill was much like the legislation the Senate had approved the previous summer. It even included the controversial section on competitive mineral leasing. The difference was in the land figures—full title to not ten but forty million acres. The administration had adopted Senator Stevens' formula: each village was to get four contiguous townships of land, or 92,160 acres, plus additional nearby land chosen for it by a single statewide development corporation. In the case of villages like Barrow, which were surrounded by an inviolable federal reserve, the village would receive surface rights to adjacent land and mineral rights to land elsewhere, selected for it by the statewide corporation, in lieu of mineral rights to reserve land surrounding the village. Wildlife refuges and ranges were to be used in the settlement although national parks and monuments were not. Neither was Pet Four.

For the most part, the state's interests were carefully protected. Senator Stevens had done his job, as he saw it, well. The Natives could not take land that had been patented to, or tentatively approved for patent to, the state. The bill also did away with the land freeze so hated by non-Native Alaskans. Instead, it withdrew twenty-five townships around each village from any entry under the public land laws for the duration of the settlement process. While this meant a partial freeze, it opened up vast areas of Alaska to immediate development once the settlement had been approved by Congress and the President.

The administration bill made other withdrawals, including the creation of a twelve-mile-wide transportation corridor along the route of the proposed pipeline. This corridor, which had first been suggested by Senator Jackson in early 1971, was opposed by the oil companies, which feared it might subject the pipeline to Congressional approval.

Although the bill was a big improvement over the handiwork of Secretary Morton's predecessors, Hickel and Udall, it did not completely please the Natives. After his meeting with the President on April 6, the day the bill was unveiled, Wright said guardedly that it was "a large step forward." "It was the first time we ever had access to the drafting of legislation," he said, reflecting the Native resentment over the closed committee meetings which had produced the 1970 Senate bill. Wright made it clear that he did not intend to give up on the AFN bill. He could not very well do otherwise. He was under intense pressure from regional groups like the ASNA not to compromise. Actually, no serious attempt was ever made in the Senate to push the AFN bill. In the House, Representative Meeds, who was sympathetic to the Natives' cause, fought in committee for portions of the AFN bill which were particularly important to the Natives, such as the regional power structure.

The importance of the administration's support of the claims cannot be underestimated. Despite its national implications, the claims settlement remained a highly specialized piece of regional legislation, not the sort upon which the White House generally takes a strong stand. But the White House plainly decided to use the Alaskan situation to back up the President's July 1970 message on Indian self-determination. The importance of the Trans Alaska Pipeline may have made a difference, but it is interesting to note that when oil lobbyists later approached the White House for support on specific amendments to the claims legislation, they were unequivocally told by Patterson that the White House wouldn't help them unless the Natives first approved it. Administration support meant a lot in Congress. When oversimplified, as it inevitably was, the legislation sounded shockingly generous to many Congressmen. The administration's support won over some reluctant Republicans. More importantly, it gave senior Republican members of the House Interior Committee guidance in writing a settlement. Later, the White House helped to round up Republican votes for the settlement when it came up for consideration in the House. Laura Bergt's kisses indeed had won several million acres for her people.

9

JUSTICE OR OIL?

". . . two percent to the justice of the case, ninety-eight percent to the need for oil."

Edward Weinberg, 1972

While the Natives were parleying with the White House, their allies, the oil companies, were putting together one of the strangest lobbying alliances Capitol Hill has ever seen. Side by side with the Natives were representatives of oil and organized labor, traditionally bitter enemies, all pushing for a "prompt and just settlement." Eventually there would have been a settlement, but it took oil to get it in 1971. One reason was that it took the oil companies' persistent interest in the claims settlement to move Representative Aspinall, chairman of the House Interior Committee. Another was that it took Alyeska's soft-spoken and sympathetic lobbyist, William C. Foster, to get the principals—the state, the Natives, and the oil companies—working together.

On the surface, Bill Foster was an unlikely ally for the Natives. He was a Washington lawyer who had corporate clients such as the Ralston Purina Company. He had extensive Alaskan connections but of the sort that would make him closer to the state than to the Natives. Foster had worked for the late Senator Bartlett and had helped to codify the new state's laws in 1959. While living in Juneau at that time, he had been a close friend of another young lawyer, John Havelock, who by 1971 was the state's Attorney General. Foster had also known Governor Egan. Although Foster was acquainted with some of the Native leaders, he did not know AFN President Don Wright.

Foster was hired early in 1971, after a meeting with Alyeska president Patton and other Alyeska executives in Washington. They asked him to draw up a legislative program. He submitted an outline of what he thought the oil companies should be doing in the claims settlement and got the job.

The position which the oil companies took, at Foster's advice, was one of simultaneously "pushing and guarding." They wanted the claims settled so that construction could start on the pipeline. But they wanted the settlement to satisfy the Natives so there would be no subsequent lawsuits to further delay their work. They also realized that the claims legislation could give the pipeline Congressional exposure that might make it possible for Congress to vote on the pipeline itself. The oil companies were afraid that they would lose such a vote, just as Boeing lost the Congressional battle over the Supersonic Transport, despite formidable lobbying for the SST by industry, labor, and the administration.

The oil lobby found itself pitted against the conservationists, who had two objectives. First, they wanted to attach a Trans Alaska Pipeline rider to the claims bill, in hopes of forcing Congress to vote on the merits of the pipeline project. Secondly, they wanted to extend the land freeze and halt any major project like the pipeline until a comprehensive land use plan could be worked out for the state, a process which was expected to take about five years. *The Living Wilderness*, the quarterly publication of the Wilderness Society, one of the plaintiffs in the suit to halt the pipeline, noted:

> Tremendous economic forces are now poised, ready to lay waste to Alaska once a Native claims settlement becomes law and the

> great public domain is opened to all comers. The only way to meet this threat and assure a fair settlement for the Natives, we are convinced, is to provide for a proper master plan. Without such a plan there can be no real protection of Alaska's resources for either the Native peoples or the 200 million other Americans who share in Alaska's vast public ownership.[104]

As long as the conservationists' land use plan did not interfere with their getting the land they wanted, most Natives were pragmatic about them. But they had few illusions. "Don't worry about the conservationists," Charlie Edwardsen told BP lobbyist Hugh Gallagher about this time. "We're playing them like a trout."

Gallagher, who was, with Foster, one of the architects of the oil-labor-Native coalition, later said their massive lobbying effort to touch all bases might have been "overkill," but added that there was always an unknown factor—what the conservationists would do. Actually the conservationists were rather less effective than they might have been in this instance, because they had to work through a rather irrational ally, Representative Saylor. "Conservationists scare Congress more than they really should," remarked Gallagher when it was all over, but in 1971 he was taking no chances.

Land use planning had advocates in Alaska as well as in the lower forty-eight. A small but influential group of Alaskans had long wanted a planning commission for the new state. For the most part they were not conservationists but economists and planners, like Victor Fischer of the university's Institute of Social, Economic and Government Research, and Joseph H. Fitzgerald, once head of the Federal Field Committee for Development Planning in Alaska and now employed by Atlantic-Richfield in Alaska. Fischer, Fitzgerald, and others saw Alaska as an excellent testing ground for economic and social planning, perhaps the last place in America where planning could still precede development. The Field Committee had done some work of this sort but its legislative authority was to expire in the spring of 1971. Through Douglas

Jones, an economist who had worked with Fitzgerald on the massive 1968 Field Committee study of the land claims, the planning advocates persuaded Senator Gravel to introduce legislation to establish some sort of land use planning commission for Alaska.

In mid-March, only a few days after Stewart M. Brandborg, the Wilderness Society's Washington lobbyist, had asked the Senate Interior Committee to consider attaching a five-year land use planning program for Alaska to the claims bill, Senator Gravel proposed a federal-state land use planning commission with broad and flexible powers. The proposal was a separate piece of legislation but Senator Gravel meant it to be considered in the context of the claims settlement. The principal difference between the Alaskan's proposed commission and the Wilderness Society's five-year program was that the commission would not delay the pipeline's progress. In May, the Alaska legislature approved state participation in the commission. Meanwhile, in the House, Representative Saylor let it be known that he would try to amend any claims settlement legislation to require comprehensive land use planning that would halt the pipeline.

With conservationist opposition to the settlement shaping up, Foster got to work quickly. The first problem was House Interior Committee Chairman Aspinall. Aspinall had started out denying that the claims had any validity at all. Foster talked to him. Gallagher, who had once worked for his Committee, talked to him. Governor Egan, an old friend from the days of the statehood battle, talked to him. All stressed two things: the importance of the settlement to the state's future and its relationship to the impending energy crisis. As an architect of Alaskan statehood, Representative Aspinall tended to regard the claims, despite his doubts about their legality, as something left over from statehood which it was his duty to settle. But he was also worried about the developing energy crisis, according to Foster, and it was because of this that he eventually yielded ground on his stand on the claims. Once convinced of the importance of one to the other, Representative

Aspinall was willing to put extra time into getting the claims bill out of his Committee immediately. To win Representative Aspinall over, the oil companies had to be unified, and until Alyeska hired Foster they were not unified. As one close observer noted, the companies had been "bumping into each other" in a variety of ineffective attempts to expedite the claims bill or else had been doing nothing at all.

Foster also approached the other patriarchs of the Committee, Representatives Haley and Saylor. But Haley meant what he had said earlier about having nothing to do with lobbyists. To the very end, the only outside people he would talk to about the claims were Governor Egan and Senator Bartlett's widow, Vide. Representative Saylor was willing to see lobbyists but Foster did not get very far with him either. He even arranged for representatives of Bethlehem Steel, whose interests Representative Saylor had indicated he was protecting, to talk to the Congressman. The Bethlehem Steel men took Representative Saylor out to dinner but he refused to explain his objections to the pipeline to them. They were mystified.

Meanwhile, during these maneuvers by the oilmen, the Natives and their chief spokesman in the House, Representative Meeds, were planning to ride over Chairman Aspinall. There was a slim possibility that they could do this because the Reorganization Act of 1970 and a revolt against the Chairman's dictatorial ways, staged by Representative Meeds and other liberals, had changed the House Interior Committee's rules slightly. The new Committee rules did eventually play a role in getting the claims bill through the Committee, but they did not endear the rebels to Representative Aspinall. Representative Meeds, an articulate liberal who was naturally sympathetic to minority causes and who had many Indians, even a few Alaska Natives, living in his district in Washington state, worked hard for the AFN. He had his own staff working on legislative proposals even though only a tiny minority of his constituents could expect to get anything out of the settlement.

Representative Meeds was also a friend of Foster, who had

once managed one of his campaigns for Congress. Foster tried to convince the Congressman that he was butting his head against a brick wall by tackling Chairman Aspinall head on, but Meeds did not see it that way. He remained determined to force the claims bill through the Committee, despite the Chairman.

In early 1971, the Natives and the oil companies were not working together. "I used to see them from time to time," said Foster. "It used to bother me. I was retained to help the Natives, but I was not that close to them." The Natives were busy with the White House and their own lobbying. There was little communication between allies. Hugh Gallagher of BP remained the principal contact between oil and the Natives because of his friendship with Edwardsen, for which he was criticized by several Washington representatives of Alyeska's parent companies. (One used to refer to the two men as "the Frito Banditos.") But the men remained friends and Edwardsen continued to use BP Alaska's plush Washington office as one of his centers of operation.

Soon after he was retained by Alyeska, Foster decided that he would need the help of organized labor, particularly maritime labor, which had a stake in the pipeline project because huge new tankers would be needed to carry the oil south from Valdez. In an early conversation with Foster, Senator Stevens, a member of the Senate Commerce Committee, which handles maritime matters, suggested the Alyeska lobbyist approach the Seafarers International Union (SIU). Foster went to talk with William Moody, the SIU's Washington lobbyist. Foster also brought Gallagher into the act since BP would be moving about 60 percent of the oil but did not have its own fleet. The two men approached Paul Hall, the SIU's colorful president and a friend of the late Senator Bartlett, for whom both Foster and Gallagher had worked.

After discussions with Moody, Hall sent word that he was interested but wanted to know what was in it for his union. Tankers, they told him, built in American shipyards and manned by American crews. But Hall was cautious and shrewd. He wanted assurances that only members of his union would benefit, assurances which the oil companies were obviously in no position to give. Hall suggested a meeting with his corporate counterparts, the presidents of Alyeska's parent companies. The oilmen's lawyers advised them not to attend such a meeting.

The negotiations between the SIU and the oil companies seemed to be at an impasse when Foster and Gallagher had a brainstorm. Since BP had the most to offer maritime labor they set up a meeting between the SIU president and Frank Rickwood, president of BP Alaska, Inc., a small subsidiary of the mother company, whose president was not at all what Hall had meant by his "counterpart." Rickwood rented a private club in New York, ordered the fanciest meal he could get, and gave a luncheon for Hall in the spring of 1971.

To Foster's and Gallagher's surprise, Hall came, accompanied by a bevy of lawyers and advisers, took one scotch, allowed himself one comment on the business at hand—something to the effect that the SIU had never held up a ship—and spent the rest of the afternoon telling stories about Captain Cook, for whom Cook Inlet in Alaska is named. Afterward, Hall's host was discouraged. Hall was a fascinating raconteur and a charming man, Rickwood complained to Gallagher, but surely the meeting had been a failure since there had been no discussion of business. "Don't worry," Gallagher assured him. "It was remarkable that he came and even more remarkable that he stayed."

That was Friday. Monday morning, SIU lobbyist Moody called Foster and said, "Paul says it's OK. 'Do anything Foster wants.' " This led to an SIU resolution endorsing the pipeline, telegrams from Hall to all members of the House Interior Committee about the claims settlement, and other lobbying assistance. Hall was also influential in getting the AFL-CIO to endorse both the pipeline and the claims settlement. However, Representative Meeds, a member of the House Education and Labor Committee as well as the Interior Committee, also had a hand in this. Through his good contacts with labor, Representative Meeds was eventually able to persuade AFL-CIO president George

Meany to write all members of the House in support of a generous claims settlement. And this was only the tip of the iceberg. During the crucial week or so before the claims bill came up on the House floor, when the conservationist opponents were most active, no less than fourteen labor lobbyists made the rounds of the House office buildings.

In addition to the aid received from the oil companies and from organized labor, the Natives got some help from civil rights groups. The Natives had always been reluctant to enlist Black support for their cause. They were not sure they wanted to return the favor and, for the most part, they did not approve of Black militants' tactics. But most importantly, the Blacks were welfare-oriented, in the Natives' opinion. "Our case is real estate," said Wright. However, the Natives' white allies did not always look at things that way. They tended to view the Natives as a minority seeking justice, like Blacks or Chicanos. Representative Begich met with Clarence Mitchell, the NAACP's skillful Washington lobbyist, who also represents the Leadership Conference on Civil Rights, and asked him to line up support. On the morning of the day the claims bill came up on the House floor, every Congressman received a telegram urging him to support the Committee bill and signed by 125 civil rights groups.

Foster also tried to bring the state into the lobbying alliance, but the state was always suspicious of the Natives (and vice versa), despite the era of good feelings initiated by Governor Egan and Havelock. At one point, Wright and several other Native leaders were in Juneau, circulating a petition in support of the AFN bill among members of the legislature. (The petition was later presented to Congress.) Most legislators signed the petition, which called for 60 million acres for the Natives, and then denied that they had understood what the petition said. They said Wright had misrepresented it to them and withdrew their endorsement. This activity kept Governor Egan, whose office was in the same building as the legislature's chambers, on pins and needles. He kept wondering what was going on downstairs and suspected Wright of trying to undermine him.

As the curious oil-Native-labor alliance was being pieced together, the Natives appeared before the Senate Interior Committee a final time. The Committee wanted to know what the Natives thought about the administration's new bill. The discussion centered around Senator Stevens' maps and overlays. The Natives pointed out that, as the bill called for the Natives to choose land contiguous to their villages, and as many villages were surrounded by state land, there would have to be negotiations between the Natives and the state. They wanted to sit in on the Committee's mark-up sessions.

The Natives also wanted the state in on the sessions. AFN attorney Weinberg said:

> In effect, this is a real estate transaction that is involved here. Everybody knows it, the committee knows it, I know it, the oil companies know it, the state of Alaska knows it. But it is very difficult to hammer out a real estate transaction in the formalities of sitting at a witness table with the committee sitting in back and with the state of Alaska which has an interest in this matter not even present and also not able to participate in the discussion. They are as much inconvenienced and discommoded by the rigid form in which we must carry on what should be the subject of negotiations as are we and as is the committee. So there is something to be said for everybody getting into a room and trying to hammer out an agreement in the give and take of negotiation.[105]

Governor Egan did not appear, nor did any non-Native Alaskan other than a spokesman for the State Chamber of Commerce, who said that this group was "softening." At the close of the meeting, Senator Stevens, who was presiding, asked a banker from Fairbanks who had been listening all day in the back of the room if he had any thoughts. "If the Natives get all this land," said G. W. Stroecker, "there won't be any acreage left to homestead or t & m [trade and manufacturing] sites or other types of land acquisition by the individual [*sic*]. So what I am concerned with is what is the future for the individual Alaskan, his children, and for the individuals who will be coming to Alaska in the future?" [106]

On the other side of Capitol Hill, the problem was at once more simple and more complex. Should there be a settlement? If so, what form should it take?

House Interior Committee Chairman Aspinall had been waiting to see what the administration would do. Once Secretary Morton sent up a bill, he arranged for Representative Haley's Indian Affairs Subcommittee to hold hearings on it. Like Aspinall, Governor Egan was waiting to see what the administration would do and now he agreed to come to Washington on May 4 to comment on the new bill. It was an event Alaskans, like Congress, awaited anxiously since the Governor had been evasive about specifics in the past.

The Governor told the subcommittee the state was willing to contribute land that it might otherwise be able to select, provided the federal government contributed twice as much land from its own reserves. The state was willing to provide up to 20 million acres if the federal government would provide 40 million; this would add up to the 60 million figure the AFN favored. Why not use the 23 million acres in Pet Four, the Governor suggested, and another 17 million from parks and wildlife refuges in the state? Politically, the Governor's move was a clever one since it made him appear to be extremely generous. Actually, since there was little chance that Pet Four could be used, the Governor was still refusing to be pinned down to a position on the claims.

The subcommittee kept the Governor on the witness stand all day, while Native leaders waited restlessly. They barely had a chance to start their statements when Representative Haley asked them if they could come back in the morning. They said they could. Wright and other Native leaders, representing the AFN, testified all day Wednesday and much of Thursday. The Committee questioned them minutely and listened to lengthy explanations and legal arguments. At the end, Representative Haley looked down at

Wright and said, "Mr. Wright, let me ask you this: We have sat here for four days. Do you think that you have had an opportunity to fully present the views of your people?"

"Yes," replied Wright, "I think this is the best hearing I have ever attended in my life." [107]

Toward the end of the second day of his testimony, Wright was asked about the differences between the AFN and the ASNA by committee counsel Lewis Sigler. Did Wright know that Joe Upicksoun, executive director of the ASNA, had submitted a statement which outlined a totally different settlement than the one the AFN had proposed, Sigler asked. Wright hedged, obviously sensing difficulty. "I am not aware of that particular testimony, no." Later he suggested that the statement contained Upicksoun's "personal views." Sigler did not pursue the point. "We will ask Mr. Upicksoun about it," he said.

When Upicksoun, an intense young man from Barrow, took the witness stand on Friday morning, Representative Haley insisted that he read his long printed statement, although it had been submitted to the Committee several days before. Representatives Aspinall and Haley "should be retired from public life," said Upicksoun, who was accompanied by his lawyer and Edwardsen. "They are unworthy." Upicksoun said the Natives were angry about a memorandum he said had been written for Chairman Aspinall by the Committee staff in December 1970. According to Upicksoun, the memorandum said: "The 40 million acre figure is used for psychological purposes. It could be omitted and the Secretary would then grant subsistence permits on whatever acreage is needed." [108] The memorandum referred to the legislation Representative Aspinall had introduced in 1971. The Congressmen were trying to dupe the Natives, Upicksoun said angrily.

He was in the middle of his statement, a little embarrassed by having to read it after Representative Haley's long and solicitous hearings, when Representative Roy Taylor, D-N.C., a courtly southerner, interrupted. "You have no power to retire these men from public office and you are in no position to determine whether or not they are unworthy," he scolded. (Edwardsen, who was sitting beside Upicksoun that day, later made several speeches in Representative Aspinall's Colorado district before the 1972 primary, in which the Chairman was defeated.)

Even Representative Meeds, usually sympathetic to the Natives, rebuked Upicksoun sharply for a counterproductive "breach of the decorum of this committee." Upicksoun tried to apologize to the Chairman. "No apologies necessary," said the Congressman. "The statement stands and you have had plenty of time to think it over." [109] When the hearings ended, everyone was apprehensive about how much damage Upicksoun had done. The Natives and the Congressional delegation remembered how angrily Representative Haley had reacted in 1970 when his feelings were hurt.

The incident caused some delay. Representative Haley, who had said he would hold mark-up sessions immediately after the public hearings, postponed them indefinitely. Representative Aspinall told the *Oil Daily* that conflicting claims by Natives and last-minute policy shifts by the Interior Department could delay the legislation for months. Senator Jackson, who had decided to let the House take the initiative, did nothing.

In Alaska, the spring of 1971 was a time of fear. The *Fairbanks News-Miner* succinctly summed up the response to the administration bill, "Non-Natives have rights, too." [110] What had happened was that the Alaskan establishment had come around grudgingly to accepting the Natives' two percent royalty as part of the cost of getting the North Slope oil to market. But President Nixon's endorsement of 40 million acres for the Natives set off a new wave of anti-Native sentiment. And the land question touched people who had not been particularly concerned about the royalty, like homesteaders. Some had trouble understanding why the Natives wanted so much land. "Over the years I have watched the Native land claims grow and grow," a man from the Kenai Peninsula wrote a member of the Congressional delegation. "At first I was very much in favor of a land claims settlement [by the federal government] but now, like many others, I am beginning to wonder.

They say they need the money to catch up with the 20th Century and, at the same time, the land so they can go back to living like they did 200 years ago."

Nor had the special interests backed down an inch. Miners and businessmen worried about an extended land freeze. The Professional Hunters Association saw the Natives snatching up 60 percent of all the game. Local conservationists had visions of dismembered parks and wildlife refuges. Homesteaders worried about their claims. As the Natives became increasingly self-confident and militant, white Alaska built around itself its own barrier of fear.

The special interests turned to Senator Stevens. While the individuals and associations wired and wrote all three members of Alaska's Congressional delegation, it was Senator Stevens, who they hoped would save them from the claims settlement. He was a Republican with a record of taking good care of his constituents' problems; perhaps more importantly, he was one of them, a former attorney whose clients had included oil companies and other businesses (although Natives, too). The Senator had to stand for reelection in 1972 when the late Senator Bartlett's term expired. He promised to look out for his constituents, but many seem to have been unsure that he could or would, even if he did go back to Alaska almost every other weekend.

That spring, the Senator received a letter from an attorney who had been a political ally. "I think you should know that the people I have talked to are up in arms over the native claims bill," this lawyer wrote. "No one whom I have spoken with favors these bills and most of them are dead set against them. . . . As you know . . . I have long supported you and I think that you have done a fine job in Washington but I feel that you are on shaky grounds on the native claims bills."

"I am quite aware that your letter . . . accurately reflects the sentiment that exists among a large portion of the Alaskan public," replied the Senator. "However, as a lawyer, I am certain you have seen the reactions of people to demands made in personal injury or property damage suits. We are still in that situation. In order to get a bill that is fair to everyone, in the end, I am sure you realize I have the same position as an attorney in determining how much to put in the demand for damages." And he appended a handwritten note. "We've got to get that bill passed. It's going to destroy us if we don't do it. I know you hear the comments I heard . . . but I've got to try to get a fair bill passed."

Senator Gravel was in a different position. While he publicly supported the AFN bill, he clearly had his own notion of what a settlement ought to be, a notion which was strongly influenced by the thinking of the Federal Field Committee. The Committee had tried, in good faith, to work out a settlement which would be in the best interests of both the Natives and the state, as well as a statewide settlement which would resolve potential conflicts among the Natives themselves. The Committee wanted to give the Natives little land but a stake in the state's economic future. It also sought to avoid creating numerous remote racially based enclaves. But, in 1971, the administration, the Natives, the state, and Congress seemed headed in the opposite direction, toward a large land settlement with the emphasis on regional autonomy. This made Senator Gravel the odd man out. Alyeska lobbyist Foster remembers having lunch with him during this period and trying to convince him of the reasonableness of what the Natives wanted. "He was getting good professional advice that 40 million acres and regional corporations might not be the best public policy," he said. "Mike was sensitive to this thinking and it took him some time to come around to what the Natives wanted." The Senator also had his own pet interests, among them the land use planning commission. The commission was eventually included in the settlement legislation.

Senator Gravel also had problems which had nothing to do with the Native claims. These were largely the result of his anti–Vietnam War activities. Senator Gravel had moved from the middle-of-the-road stance he had taken when he defeated Senator Gruening, one of the Senate's earliest and staunchest doves, to a rather strong antiwar position. By the spring of 1971, he was embarked upon a filibuster to prevent Congress from extending the Selective Service Act, since he believed that without the draft the Vietnam War would come to an end. Only a short time later, he was to read some of the Pentagon Papers at an all-night session of a minor subcommittee, a performance which ended with the Alaskan in tears. The Senator had gotten badly out of step with his constituents, who were mostly a hawkish bunch. Furthermore, many of them, including many Natives, felt that his job was to solve Alaska's peculiar problem, not to tackle America's.

Representative Begich, a moderate-to-liberal Democrat, was initially suspect to the conservative Alaska business establishment. However, his assiduous refusal to be backed into a corner on the claims by the Natives and his kid-glove handling of Representative Aspinall won their admiration. The Alaskan establishment, from the *Anchorage Daily Times* on down, allocated a specific role to Begich, one which he played to perfection. His job was to get a bill, any bill, through the House and into conference with the Senate, the very thing his predecessor, Howard Pollock, had been unable to do.

From April to mid-June, Alaska was a cauldron of simmering hatred and fear. Everyone waited for Congress to make a move. Finally, Representative Haley arranged for six days of mark-up sessions in late June. On Monday, June 21, twelve of the fifteen members of the Indian Affairs Subcommittee met on the third floor of the Longworth Building. (The Chairman and the ranking minority member of the full Committee are ex-officio but voting members of all the subcommittees.) Representative Meeds was there to represent the Natives; Representative Begich, the state; Representative Sam Steiger, R-Ariz., and John H. Kyl, R-Iowa, the administration; and Representative Saylor, the conservationists. All were in touch with the oil lobbyists. Representative Meeds, still trying to run over House Interior Committee Chairman Aspinall, had put together a coalition with the Republican subcommittee members which seemed to have the votes to do just that. First, the coalition voted down a new bill Representative Aspinall had had attorney Sigler draft in June, which gave the Natives a little more land in fee and a little less for subsistence use. Then the coalition voted to take up the administration bill and the subcommittee began to vote on specific parts of it, starting with the least substantive. The conclusive vote came on a relatively unimportant matter, but one about which the Chairman felt strongly. On July 7, the subcommittee voted on whether there should be an Alaska Native Commission to administer the settlement. The Chairman opposed the commission. The vote was nine to eight. The Chairman lost. After the vote, he claimed he heard bells ringing calling Congressmen to the floor, and Representative Haley adjourned the meeting and would not set a date for another session. It looked as though the subcommittee would not meet again before Congress recessed for the summer.

Representative Aspinall had seen the writing on the wall. Fearing the coalition could pass a bill over his objections, he did not want the subcommittee to meet again. Representative Haley, who was more or less his cipher, agreed to stall, although both men knew that under the new rules subcommittee members could petition and call another meeting themselves.

Right after this abortive meeting, Foster got a call from Don Wright, asking if he and other Native leaders could come over and talk. Wright arrived a little later with several associates, including State Senator Willie Hensley and Charlie Edwardsen. "OK," said Wright to Alyeska's lobbyist, "what do you think?"

"You can't ride over Aspinall four times in a row—in subcommittee, in committee, on the floor, and in conference," Foster told the Natives. "Once maybe, but not four times. So go with him, not against him."

This was a bitter pill for the Natives. It took a lot of talking by Foster to convince them of the reasonableness of his advice. He pointed out that the time was ripe for a generous settlement but it might not always be. If the Natives prolonged the fight, their chances might get worse. This meeting set the tone for a close relationship, both personal and professional, between Foster and some of the Native leaders, particularly Wright. Foster also managed to maintain good relations with Edwardsen, whose belligerence often turned off potential allies. Before crucial votes, Foster made a practice of consulting both Wright and Edwardsen to see what their head counts were.

Meanwhile, Representative Begich devoted the month of July to winning over Chairman Aspinall. Fortified by having voted with him on the Alaska Native Commission, Representative Begich went to see the old man. The Chairman was courteous but vague. He indicated he might be willing to try again if the members of the subcommittee would agree in advance to certain compromises. Then he told Representative Begich "he and Haley" had been thinking and felt that perhaps there were more dimensions to the settlement than just space. There was also time. Might it be possible to give the Natives some land now and more later? He left the Alaska Congressman with this hint.

Representative Begich went to see Representative Haley, who was equally vague. He also talked with Representatives Meeds, Kyl, Steiger, and Edmondson. There was talk of forcing a meeting of the subcommittee.

By now, it was late July. Congress was expected to start its vacation on August 6. Time was running out. Representative Begich went to see the Chairman again. This time, Representative Aspinall produced a typewritten list of points he would like to see in a final bill. He told Representative Begich that if the Alaskan could get the various parties to agree on these points and initial the agreement, he would ask Representative Haley to hold another meeting of the subcommittee. Friday, July 30, was the day the House debated a $250 million loan guarantee for the Lockheed Corporation until late in the evening. Representative Begich spent the afternoon and evening on the floor, moving from Representative Kyl in one corner to Representative Meeds in the other, then over to the Chairman in the center of the floor, or into a small room on the Democratic side where Representative Haley was sitting, always with his slip of paper containing the ten points. By the end of the evening, he had the important signatures on his piece of paper and Representative Aspinall had asked Representative Haley to schedule a meeting of the subcommittee the following Tuesday. Over the weekend, Committee counsel Sigler went to work and drafted a bill based on the ten points. Meanwhile, Representative Begich contacted the other members of the subcommittee whose signatures were not on that precious slip of paper and told them what had happened, asking their forbearance.

The bill drafted by Sigler contained a number of compromises. First of all, the Congressmen had agreed to give the Natives 18 million acres around their villages immediately and another 22 million after the state had finished picking its land. The villages would have five years in which to choose their 18 million acres. Then the regional corporations might have to wait as long as twelve years until the state's authority to choose the remainder of its 102 million acres expired (perhaps longer since the authority was renewable if Congress so chose). But eventually the Natives would get full title to 40 million acres.

All mineral rights were to go to the regional corporations. These corporations were a second compromise, since Representative Aspinall opposed a single statewide organization, which he feared would grow too powerful. The twelve regional corporations were to pass on most of their assets to the villages; this arrangement was the opposite of what the Natives wanted.

The compromise gave the Natives $425 million in federal funds and a two percent royalty on all mineral revenues, a compromise with Representative Aspinall's feeling that the federal government should not contribute more than $350 million to the settlement. Instead of withdrawing all the land in the state from all entry until the villages had made their selections, the bill withdrew just the land immediately adjacent to them for five years. The bill protected the state's selections to date, regardless of the status of the patent. The Natives got no mineral rights in Pet Four. The bill protected all valid rights in land selected by the Natives: all leases, mineral claims, and homesteads. And it contained neither a pipeline corridor nor land use planning provisions.

On Tuesday, August 3, the subcommittee met and Representative Begich formally presented Representative Haley with the ten points, all signed and in order. The only important member missing was Representative Saylor, who was in Bethesda Naval Hospital for a checkup. After seventeen minutes, the subcommittee members voted unanimously to report the bill to the full Committee. It was a great victory for Representative Begich and the oil companies and even for the Natives.

Later in August, during the recess, Representative Aspinall had a dinner party at his home, to which he invited representatives of several independent oil companies and Interior Secretary Morton. The purpose of the dinner was not to discuss the claims or the pipeline but inevitably both came up. That night, Representative Aspinall told his guests that he thought Congress would approve a claims bill by early winter.

Once the House subcommittee had acted, the Senate went to work, though not immediately through official channels. Representatives of the two Alaska Senators, attorneys for the AFN, and members of the Interior Committee staff began to meet with one another to discuss what should be in the Senate bill.

Publicly, the AFN, supported the House bill. Out of deference to the agreement with Chairman Aspinall, they and the state agreed not to try to amend it on the House floor. Actually, the Natives had a number of objections to the House bill. They still wanted all their land before the state chose any. At the very least, they preferred some sort of alternate selection process which would allow them to pick some land, then the state, then the Natives again, to the "some now, more later" approach of the House bill. They did not want the land to be tied strictly to the villages. They wanted the oil in Pet Four since there seemed little possibility that they would get any other oil land on the North Slope. And they wanted to keep 50 percent of the income earned from land in any one region within that region, a point that was particularly important to the North Slope Eskimos since much of the state's, and therefore the Natives', future revenues would come from the land the ASNA claimed.

AFN lobbyists had determined that they could raise the total land grant to 40 million acres but not 60 million if it came to a vote on the Senate floor. At this point, Senator Jackson still proposed to give them only 20 million acres. But Senator Jackson also wanted to avoid a floor fight over the acreage, fearing that he would lose control of the bill if this happened. There was a good deal of bargaining over this issue, during which AFN leaders admitted that they could not prevent the ASNA from trying to raise the acreage to 60 million on the Senate floor no matter what the Committee agreed upon. By September, the 60 million acres looked impossible and the AFN, while still talking about "free-floating selections," was willing to have about half the land tied to the villages. The AFN proposed that the Natives be given 20 to 22 million acres of village land and another 18 to 20 million acres anywhere in the state. The Natives would choose all their land before the state picked any new land.[111]

Meanwhile, the Senators' staffs were meeting to see where the two Alaskans agreed and disagreed. Once again, the two men had agreed to work together despite their personal animosity. As one lobbyist observed, "They had other ways of getting at each other." Both wanted the land tied to the villages. Both wanted most of the assets of any corporations passed on to the villages. Both favored regional corporations instead of a single statewide one, provided no one group of Natives prospered at the expense of the others. Both favored a two-tiered land selection process allowing the state to finish picking its 102 million acres before the Natives had completed their land selection and thus preserving the facade of the Statehood Act. They did not agree about land use planning, a land freeze after the settlement, and competitive mineral leasing.

Shortly before the end of the August recess, Doug Jones, representing Senator Gravel, and John Katz, representing Senator Stevens (for whom he had gone to work after Representative

Pollock's defeat), met with William J. Van Ness, the Committee's special counsel, who more or less spoke for Senator Jackson. The Chairman was willing to reconsider his 20-million-acre limit, Van Ness told them. But he felt strongly that there had to be some sort of land use planning commission in the bill to neutralize potential conservationist opposition. Jones pushed for Senator Gravel's joint federal-state commission. Katz advised Senator Stevens to reconsider the commission in view of the strength in the Senate of conservationists who wanted a comprehensive land use plan for the state attached to the settlement, a proviso which would delay the final settlement by many years. Senator Jackson also insisted that competitive mineral leasing be in the bill. He wanted to retain the elaborate corporate structure in the 1970 Senate bill to use as a bargaining tool in conference. And he wanted a transportation corridor along the pipeline route.

On September 14, the day before the Committee was to meet, the Alaskan Senators met once again to set up a game plan. Since the Committee planned to use the 1970 bill as a base, Senator Stevens would offer an amendment raising the amount of land from ten to forty million acres. In addition to the statewide corporation in the 1970 bill, the Alaskans would add seven regional corporations. They would do away with the land freeze, but give the Secretary of the Interior the right to extend it if necessary. And they would modify the section of the bill which penalized the state if it challenged the two percent royalty provision in court. The Alaskans could not agree about land use planning and the transportation corridor. And the Committee would have a chance to vote on the competitive leasing contained in the 1970 bill.

Like the House subcommittee, Senator Jackson's Committee disposed of the claims quickly—in an hour and a half. However, the meeting did not go exactly as the Alaskans had planned.

Instead of adopting the Alaskans' 40-million-acre provision, the Committee gave the Natives a choice between that and a second option, proposed by Senator Jackson, which reflected his view that the Natives should get less, but better, land. The Natives could either get 40 million acres, all of it within 25 townships of existing villages, or they could get full title to 20 million and partial title to another 10 million. The full title went to village land. The statewide corporations would get full title to two and a half million acres, to be distributed to villages which somehow came out with less than their fair share of the total land grant; surface title to two and a half million acres of timber land and two and a half million of recreation land; and subsurface title to two and a half million acres of mineral land. All 10 million acres could be selected anywhere in the state. In addition, the Natives would be allowed to use another 20 million acres of the public domain for hunting and fishing. The Natives got to vote on which distribution formula they preferred, but in either case they got all their land before the state got any. But the Natives did not like the option and Alyeska lobbyist Foster called it "Congressional abdication."

The second big issue before the Committee was competitive mineral leasing. Again the Committee compromised. Instead of eliminating Section 17 altogether, the Committee changed it so it would not be as onerous to the "independents." The compromise provided that wherever there was competitive interest in leases on federal land, the leases would have to be offered competitively. But it also allowed bidding by either bonus or a percent of the royalty, the latter option letting the small companies compete with the majors, at least in theory. In addition, the lease holder could offset the annual rent against the cost of exploration and development. Despite this carrot for the oilmen eight members of the Committee voted against the provisions, including Senator Stevens.

The Committee also established seven regional corporations in addition to the statewide ones. It gave the state the exclusive right to test the revenue sharing provisions of the bill in court, although state selections would be suspended if it did. Instead of a total freeze, the Committee withdrew 25 townships around each village from entry, except in the cases of villages on state land, where selection rights were far more limited.

The big addition to the bill was a section which had to do with land use planning and a transportation corridor. First, the Committee incorporated Senator Gravel's federal-state planning commission into the bill. Then, the committee took Senator Jackson's transportation corridor proposal and expanded it to a twelve-mile-wide North Slope Transportation and Recreation Corridor, with controlled entry into the wilderness on either side of the corridor and some insurance that facilities within the corridor would both be open to the public and compatible with the environment. This last was a large order. The Committee made it clear that this did not constitute approval of the Trans Alaska Pipeline. The oil companies had opposed the corridor because they feared additional controls conservationists might attach before the bill was finally approved. The Committee also retained a little-noticed provision in the 1970 bill which authorized the Secretary of the Interior to conduct a detailed study of public lands in Alaska that were suitable for parks, forests, or refuges, and submit his report and recommendations to Congress within three years. Lands under study would not have to be opened to entry. The final bill was not sent to the Senate floor for some time but it changed little after the September 15 Senate Interior Committee meeting.

The following day, the full House Interior Committee met to consider the claims bill the Indian Affairs Subcommittee had approved in August. One week later, on September 23, the full Committee also approved it. Although attempts were made to reduce the amount of land and money, the only serious threat to the original agreement was Representative Saylor's attempt to attach a land use planning provision to it. He wanted to require a comprehensive land use plan for the state before any development could take place on federal land there. This would have halted both pipeline construction and further state land selections. But it was defeated.

However, the Committee was beginning to feel pressure from conservation organizations and, to counteract it, adopted a substitute for Representative Saylor's amendment which had not been previously contemplated. Representative Kyl worked it out in consultation with the other principals. This provision, which the Committee unanimously approved, withdrew all unappropriated public lands from entry until the Secretary of the Interior decided that they should be reopened. But this withdrawal specifically did not apply to Native or state land selections or to the pipeline. The Kyl amendment was viewed by special interests in Alaska as an indefinite continuation of the freeze as far as private individuals were concerned. Groups like the Chambers of Commerce reacted violently to it and to the idea of 40 million acres to be given the Natives, wherever they were located.

Conservationists were busy, too. On October 5, twelve national conservation groups urged President Nixon to delay the pending claims bill until its "defects" could be corrected. "In recent months, speculators and exploiters, never identified to the public, but everywhere in evidence plying the halls of Congress, have been championing the cause of the Alaska Native claims," they wrote the President. They urged him to use his existing authority to withdraw large areas of Alaska before the bill passed.

However, the Congressional process, once set in motion, continued. On October 14, the House Rules Committee cleared the bill for the floor. The night before, representatives of more than a dozen conservation groups, led by the Wilderness Society and the Sierra Club, met and agreed upon a new land use planning amendment, which Representative Morris Udall, D-Ariz., brother of the former Secretary of the Interior, would introduce on the floor. Representative Udall was a member of the House Interior Committee, although not of the subcommittee which had drafted the claims bill. What conservationists proposed to do was give the Interior Secretary authority to set aside 50 million acres in Alaska as "a national interest study area," to be considered for inclusion in either the park system, the wildlife refuge system, the scenic rivers system, or the national forest system. The Natives could choose their first 18 million acres around their villages but any further selections made by them to round out their 40 million acres and any further selections by the state would have to wait until the Secretary had taken his 50 million acres.

At least, this is what Representative Udall said his amendment would do. He emphatically denied that it would halt construction of the pipeline. But the amendment kept changing. As finally drafted, and ironically enough one of its authors was AFN attorney Weinberg, the amendment was ambiguous. First, it withdrew all unappropriated public lands, like the Kyl amendment, at the Secretary's discretion. The Secretary was to review these lands and pick 50 million acres. But the amendment also set aside for the federal government another 50 million acres which had already been classified by the BLM as suitable for reservations, including parts of the Copper River Valley and the Brooks Range, through which the pipeline was to pass. The Secretary had five years in which to decide whether these 100 million acres should become permanent reservations. There was nothing in the amendment which specifically mentioned the pipeline and no one was sure of its effect upon the project. Representative Udall insisted that while he himself wanted pipeline construction halted until a complete environmental study of all alternatives could be made, he did not believe that his amendment would do that. He also said he did not want to take on the oil lobby just then. Other members of the Committee did not agree with his interpretation. The oil companies finally decided that the amendment could not be used to halt the pipeline, but opposed it anyway. The state and the Natives opposed it because of their agreement with Chairman Aspinall to support his Committee's compromise bill and because it gave a temporary land use planning commission the last word on land use by either the Natives or the state, as well as halting state land selections.

As October 19 approached, the day the leadership had scheduled for debate on the House floor, the conservationists intensified their lobbying. They stepped up the letter-writing campaign so that Congressmen were deluged with letters urging adoption of the Udall land use planning amendment to prevent a land rush once the claims were settled. The Native-oil-labor coalition was alarmed and set out to contact every member of the House the Monday and Tuesday before the bill came to the floor. The lobbyists coordinated their efforts by meeting daily to plot strategy. The Natives sent out their regional teams, dividing up the Congressmen's offices among them. The labor lobbyists went to see their people; oil lobbyists contacted theirs, particularly southerners. Brad Patterson at the White House started contacting Republicans, as did the Interior Department and Representative Steiger and Senator Stevens. Representative Kyl talked with Minority Leader Gerald Ford and the Republican Policy Committee. The President sent up a letter supporting the Committee bill, without amendments, which Minority Leader Ford read on the floor. The civil rights coalition sent telegrams to every member. Senator Gravel talked to Congressmen. And Representative Begich, assisted by friends on the Committee, did a massive job of personal lobbying, which won him the admiration of even his opponents, like Representative Udall.

The son of Yugoslav immigrants, Representative Begich was an ambitious, pragmatic man. He had inherited the mess created by Representative Pollock's inept handling of some difficult personalities on the House Interior Committee. Representative Begich was determined not to make the same mistakes. Known during his tenure in the Alaska legislature for his temper and ill-concealed ambition, Representative Begich changed faces in Washington. He became modest, self-effacing, and deferential when the occasion demanded it, which was most of the time. He supported Chairman Aspinall in most of his ups and downs within the Committee. He asked for the Chairman's advice on all sorts of matters. Similarly, he never rose to Representative Saylor's baiting. In fact, he went out of his way to court Representative Saylor. Once he took two Wein Consolidated Airlines pilots, both Natives, to see the ill-tempered Congressman. The two men impressed Representative Saylor very favorably. Later he frequently commented publicly upon how well they spoke English. Representative Begich sat by, smiling smugly.

The Alaskan's handling of his peers was less deferential but no less calculated. He and his legislative assistant, Guy Martin, kept charts which showed whom the Congressman had spoken with about the claims and which way they were leaning. Shortly after he arrived in Washington, Representative Begich started softening up his colleagues. About once a week, from the middle of April on, he sent Alaskan propaganda to the other House offices. One time it was a brochure about the shocking poverty in the villages. Another time it was a specially labeled can of Alaskan King Salmon. Representative Begich also used to pick out a Congressman he didn't know and sit beside him on the floor for several days in a row. They would talk about all sorts of things but eventually the Alaskan would bring up the subject of the claims. He also went to see state delegations when they met, varying his approach to suit their thinking and concerns. When he talked with southerners, he stressed states rights. "Land use planning is really a states rights issue," he said. Other southerners found the national security argument, which was being used by oil lobbyists, appealing. Said Representative Begich, "With these, you talk global politics, Arab oil and our dependency upon it." With Jewish Congressmen, Representative Begich played upon their sympathy for Israel, reasoning that it is hard for the United States to go all out for Israel when it is so heavily dependent upon Arab oil. North Slope oil would lessen that dependency, the Congressman argued. With New Englanders and midwesterners, he talked about fuel shortages, an annual problem in their chilly districts. Congressmen from the oil states and the West Coast were also receptive to pipeline arguments. As AFN attorney Weinberg remarked later, the settlement was due "two percent to the justice of the case, ninety-eight percent to the need for oil." [112]

On the afternoon of Tuesday, October 19, after nearly three hours of debate on an end-the-war amendment, the House took up the claims. There was almost no one on the floor, but Native leaders and lobbyists listened attentively in the galleries. Most of the discussion centered on the ambiguous Udall amendment, although it was not expected to come up for a vote until the next day. Representative Saylor, who was cosponsoring the Udall amendment, said that the oil interests had pushed the Committee into approving a bill that was neither equitable nor warranted. He urged his colleagues to vote against it unless the Udall amendment was attached.

In response, the Chairman talked about the Kyl substitute amendment. "The purpose of this amendment is to permit Native selections and state selections to proceed, but to stop all other dispositions of the public land, unless the Secretary determines otherwise in specific cases. This is not a land use planning provision, but it anticipates that land use planning will be authorized. Legislation providing for such land use planning is now pending before the committee . . . and some hearings have been held." [113] (Three years later, Congress had still passed no land use planning legislation.) Representative Aspinall then said that his Committee had been urged to expand the Kyl provision to freeze all public land transactions other than Native selections until a state land use plan had been drawn up and approved by Congress. The purpose would be to stop further state land selections and the Trans Alaska Pipeline, he said. The Committee had decided against this course.

There were a variety of questions from the floor. A member of the Armed Services Committee wanted assurances that the Natives would not be able to get their hands on the oil in Pet Four. The Congressman from Seattle wanted to know if there was anything in the Committee bill which would further delay the delivery of North Slope oil to his city. Someone wondered if the Udall amendment would not halt the pipeline. "It has nothing to do with the pipeline," protested Representative Udall.

Representative Saylor then urged fellow Congressmen to "expand upon the planning already in the bill" by approving the Udall amendment. This was an important point because whoever was presiding over the debate when the Udall amendment came up for consideration would have to decide whether or not it was germane. Representative Udall was not sure it was. It depended upon whether the Kyl provisions already in the bill were considered land use planning provisions.

By prearrangement, someone called for a quorum. During the quorum call, Representative Begich used the Committee's maps, which were on display in the well of the House, to illustrate a brief lecture on the claims. "It was just like holding school," said Representative Begich, a former schoolteacher. Forty or fifty Congressmen gathered to listen.

After the roll had been completed, Representative Udall took the floor to discuss his amendment. "My amendment does not slow down the pipeline in any way, shape or form," he said. "You can be for the pipeline and this amendment with perfect consistency." [114] Representative Steiger pointed out that three areas to be withdrawn by the amendment lay in the pipeline's path. "I have asked the best lawyers I can find and the staff of our committee if anything in my proposed amendment would prevent the Secretary of the Interior from approving any pipeline tomorrow just as quickly as he could approve it today and they say there is not," replied Representative Udall.[115]

The discussion continued, although there were few members on the floor. At the end of the day, Representative Begich had his turn. The Alaskan freshman had been advised by Chairman Aspinall not to speak too soon in the debate. "You fail or succeed based on what you do in the well," Representative Begich recalled the veteran Congressman saying to him. He said Representative Aspinall told him a story about former Representative Roman C. Pucinski, D-Ill., who, when he first came to Congress, talked too often and too long in the well. Pretty soon, Representative Aspinall said, no one took Representative Pucinski seriously. Therefore, Representative Begich's own remarks were general, brief, and to the point. After he had finished, the House adjourned for the day.

On Wednesday afternoon, the House considered a number of

minor amendments, accepting the technical ones proposed by Representative Aspinall and rejecting the others. Then the clerk read the Udall amendment. Representative Aspinall immediately made a point of order that it was not germane. "It is said that the intent of this legislation is very narrow," retorted the amendment's chief sponsor, "and that we are going to settle only Native Indian claims, but the fact of the matter is on the back of this legislation rides the whole future of Alaska. . . ." [116]

Representative William H. Natcher, D-Ky., who was in the chair, ruled that the amendment was germane, but the situation was immediately muddied by a clarifying amendment proposed by Representative Elford Cederberg, R-Mich., stipulating that no land could be withdrawn to block the pipeline. Alyeska lobbyist Foster was out in the hall talking to someone when the Michigan Congressman introduced his amendment. He returned to the gallery and a labor lobbyist caught his eye. "What do we do about this?" he wanted to know. Foster didn't know what he was talking about. The Cederberg amendment had been totally unexpected and, before any of the lobbyists could do anything, Representatives Saylor and Udall had accepted it as part of their amendment.

As far as the Natives and the state of Alaska were concerned, the major objection to Representative Udall's proposal had nothing to do with the pipeline. The amendment would have allowed the federal government to set aside park land before the state and the Natives had a chance to pick any land. Furthermore, it would have required that both consult a temporary land use planning commission before doing anything with their land. Because the Udall amendment was so poorly drafted, it was unclear exactly what the authority of this commission would be, but it appeared that neither the state nor a Native village would be able to sell or develop its land without first consulting the commission. Led by Foster, the oil companies continued to oppose the Udall amendment, even after they had decided that it would not halt pipeline construction, because of what it would do to the state and the Natives. But Foster feared that Republicans would support the

Udall amendment once assured by Representative Cederberg's addition that it would not affect the pipeline adversely. Foster grew even more apprehensive when a Washington State Republican named Thomas M. Pelly rose to say that "a lot of us" could vote for the Udall amendment once the Cederberg proviso had been attached.

Representative Begich objected. "This is very fine," he said, "but it conceals the real impact of the amendment on the Natives of Alaska and the state of Alaska. . . . Would you be willing to have it done to your own state?" The Udall amendment would delay Alaska's economic recovery for years and hurt the Natives every bit as much as other residents, or perhaps more, he said. "It is an amendment grounded in sound environmental philosophy and commitment, but it is an amendment which is roughhewn and inequitable in its execution, one which thoughtlessly ignores other crucial responsibilities," he said.[117]

Representative Meeds, a personal friend of Representative Udall, angrily charged that the amendment would frustrate the Natives' settlement. There would be no good land left by the time the Natives got around to their second set of selections, he charged. And what about the provision that they would have to consult the planning commission on how to use the land they got? Just "friendly advice" which they could reject if they wished, replied Representative Udall. His friend from Washington was not convinced.

At this point, a black Congressman, Representative Ron Dellums, D-Calif., pointed out that the only just way to handle the situation was to give the Natives their 40 million acres off the top with no strings attached. If the Alaska Natives had already agreed to the two-part selection process, then there was nothing anti-Native about the Udall amendment, he said. He wanted to know whether the Natives had really approved such an arrangement. Representative Begich hastened across the floor to assure him privately that they had. "It is not my amendment that is anti-Native," said Representative Udall while this conference was going on, "it is the committee bill which the Natives agreed to. I think they have been had. I think they were poorly advised. But plainly, with the advice of their lawyers, they have decided to go into it."[118]

The vote was close—217 to 177. The Udall amendment was defeated by only forty votes. The bill remained as the Committee had reported it. There was one more conservation amendment, having to do with the seal harvest on the Pribilof Islands, which was rejected. Then came the final vote on the bill itself. The claims settlement agreed upon by Representative Aspinall, Representative Begich, and other principals in August was approved by the House of Representatives—334 to 63. The "no" votes included Representative Saylor; fiscal conservatives like Representatives H. R. Gross, R-Iowa, and Durwood Hall, R-Mo.; conservative Republicans like Representatives John Ashbrook of Ohio and Joel Broyhill of Virginia; and conservative Democrats like John Rarick of Louisiana. But Democrats like Neal Smith of Iowa, whose remarks on the floor made it clear that he did not think the Natives were getting enough, also voted against the bill. As he had promised to do earlier, Representative Udall voted for the legislation regardless of the fate of his conservationist amendment.

A few days more than 104 years after the purchase of Alaska, the House passed settlement legislation which, while it did not please the Natives completely, went a long way toward settling the land claims equitably. In retrospect, there is little doubt that oil made it possible. Because of oil, Chairman Aspinall was able to compromise on the settlement. The Chairman made a 180-degree shift in point of view once the companies had made clear to him how they felt about the bill. It was not really a matter of being bought off by the oil companies, although many of their executives had contributed to his past campaigns. Representative Aspinall's traditional western views on the development of the public lands coincided with the companies' interests in this instance.

On the other hand, the conservationists bungled the claims bill in the House, although they were able to recoup their losses later. Their first mistake was in working through Representative Saylor, whose technique was obstruction and uneven obstruction at that. As an opponent in the claims fight noted, Representative Saylor was "irrational, cunning and devious." But he was also erratic. Furthermore, another champion might have participated in the agreement which led to the final Committee bill, basically the bill that the House passed. Representative Saylor was unable to do so. Later, the conservationists turned to Representative Udall, a capable and thoughtful legislator with considerable influence in the House. Although Representative Udall lost on the floor, he became a member of the conference committee which drafted the final legislation and there he drove a hard bargain for the conservationists.

10
THE DELICATE COMPROMISE

With almost no discussion and no floor fights, the Senate passed a version of the Native claims legislation on Monday, November 1, 1971. This was possible because Senator Jackson and Senator Alan Bible, D-Nev., representing the conservationists, had agreed about where the Natives' land should be. Senator Jackson was afraid that an attempt to change the acreage in either the Natives' or the state's land options would lead to restrictive conservation and land use amendments. So the conservationists agreed to the land specified in the bill approved in September by Senator Jackson's Committee, land located mostly around the villages, and Senator Jackson accepted a rather mild conservation amendment which allowed the Secretary of the Interior to withdraw land for possible inclusion in one of the park or refuge systems.

Earlier in the fall, Senators Mike Gravel and George McGovern, D-S.D., had intended to offer an amendment giving the Natives full title to 40 million acres under the Jackson option as well as the other option in the bill. (The so-called Jackson option gave them full title to only 20 million acres around the villages and various kinds of partial title to ten million acres scattered across the state, while the other option gave them full title to 40 million acres, largely contiguous to the villages.) What had happened was that, on October 4, just before the AFN's annual convention, Senator Gravel wrote its executive director, Harry Carter, and offered to do whatever the Natives wanted him to do when the bill came to the floor. One possibility, he said, was to amend the bill so that the Natives would like it better. About ten days after that, the Senator met with AFN attorneys and they told him their clients would like to get full title to 40 million acres whichever way the land was distributed. A few days later, Senator McGovern notified Senator Jackson that he would propose an AFN amendment doing just this and, on October 19, Senator Gravel told Senator Jackson the same thing. However, when the bill was reported out of committee on October 21, the two land options were the same as when first conceived. It turned out that Senators McGovern and Gravel had agreed not to amend the bill in committee or on the floor on the theory that changes could be made in conference with the House. On the floor, Senator McGovern asked Senator Jackson if the land provisions could be improved in conference.

"These options give us a lot of leeway," replied the Chairman cautiously, "so that we have an opportunity to improve upon the status of the land provisions of the bill. . . . I regret that I cannot take to conference any amendments of this kind, because with a roll call vote we may tie our hands in conference, and I do not want to do that. I can only assure the Senate that I will go to conference with the idea in mind of trying to improve upon the Senate version of the bill. . . ." [119] It was hardly a clear statement that he would try to do what the AFN wanted but Senator Gravel chose to interpret it as such. Senator Gravel told the Senate, "I want to make clear that we will be pursuing a goal in conference and that it will be a 30–10 goal in fee [a reference to the AFN plan to give the villages 30 million acres and the statewide corporation 10 million], which I think would be a very adequate settlement." He continued, "Certainly it does not bind any member of the committee or the conference, but it does give testimony that a segment of the conference—a sizeable segment—under the leadership of the junior senator from Washington [Jackson] will be pursuing very vigorously in conference this very proper goal." [120] It was all rhetoric. That was the end of the AFN amendment.

The Senate then took up an expansion of the land use planning sections of the bill along lines proposed by Senator Bible. After the Senate Interior Committee had combined Senator Gravel's land use planning commission and Senator Jackson's transportation corridor and given the Interior Secretary authority to set aside land for possible inclusion in one of the national conservation systems, conservationists had raised legitimate questions about how the whole process would work. At their urging, Senator Bible had had a clarifying amendment drafted. It required the Secretary to report to Congress every six months for three years what land he wanted included in park or refuge systems. All recommended land was withdrawn for five years, to give Congress time to act on the Secretary's recommendations. During these five years, any land the state wanted that had been withdrawn could not be patented or tentatively approved for patent to the state. However, Native villages were to receive their patents immediately (there was no mention of the Native corporations). The proposal was entirely prospective; it did not affect previously selected land like Prudhoe Bay.

The state wasn't happy with this. "If I had my druthers, I would not have them in the bill," said Senator Stevens glumly of the Bible clarifications. But, bowing to the conservationists' strength, he did not oppose them. The Bible amendment was adopted by voice vote.

Next came a series of amendments by the Alaskan Senators, including one introduced by Senator Stevens at the insistence of the Alaska Miners Association. During the land freeze, prospectors had continued to locate minerals, although they could not patent their claims. Now, with the end of the freeze in sight, they wanted to be able to establish their claims before the federal government opened the land to entry. Senator Stevens had intended to introduce the amendment as the miners had drafted it, but Senator

Jackson persuaded him to modify it so that a prospector could file his claim but would have no right to it until the Interior Secretary had classified the land for mineral use. The amendment was agreed upon, although the miners continued to work hard to get their original version into the final bill.

There were other Alaskan amendments, which were favorable to the Natives. Villages inadvertently left off the list of those eligible for land were added. Certain Tsimshian Indians were included in the settlement. An amendment provided that the settlement would not terminate any federal loan or grant programs in the state. The amount of money to be paid the AFN for its work on the claims was raised from $350,000 to $600,000 so the AFN could repay loans made to it by other Indian groups.

Once these amendments were disposed of, Senator Howard Cannon, D-Nev., a member of the Senate Armed Services Committee, proposed preventing the Natives from getting any mineral rights in Pet Four. Under the Interior Committee bill, Natives living in Pet Four would get full title to between three and four percent of the reserve. "The Armed Services Committee has never been notified nor consulted with respect to this matter," protested Senator Cannon, who was chairman of the National Stockpile and Naval Petroleum Reserves Subcommittee. Representatives of the Navy were busy in the lobby trying to round up support for this change.

Senator Jackson replied that the Navy had been consulted and had raised no objections until the last minute. Senator Cannon wanted to know what would happen if the Natives leased their land, wells were drilled, and the producing wells started draining off the national reserves. The Secretary of the Interior presently had the authority to stop such drainage, Senator Stevens told him. But Senator Cannon wanted language worked out in conference which would explicitly protect the reserve. Senator Jackson agreed and Senator Allott, the ranking Republican on the Committee, told Cannon he personally would make sure "that the actual oil reserves of the Naval Petroleum Reserve [would not] be drained away into the hands of Native individuals. . . ." [121] Senator Cannon withdrew his amendment.

The claims bill passed the Senate by a vote of 76 to 5. Those who voted against it were all Republicans: Paul Fannin and Barry Goldwater, both of Arizona; William V. Roth, a first-term Senator from Delaware; William Saxbe, of Ohio, also a first-termer; and Strom Thurmond of South Carolina.

Once again, the Natives did not get the sort of land settlement they wanted, although the 1971 Senate bill was more liberal than the one passed the previous year. Moreover, the conservationists had succeeded in inserting a conservation section into the bill, which meant that when the House and Senate Committees met to resolve the differences between their two bills, they had a basis on which to write more stringent conservation legislation. The Senate bill did not affect pipeline construction in any way.

With Senate passage, the hardest battle still remained ahead—the House-Senate conference. Senator Jackson appointed himself, the two Alaskans, and Senators Bible, Metcalf, Allott, and Frank Church, D-Idaho, to joust with the team Representative Aspinall had selected, consisting of Representatives Haley, Saylor, Udall, Edmondson, Kyl, Begich, Steiger, Meeds, and John N. Camp, R-Okla. Representatives Meeds and, to a lesser degree, Edmondson, could be expected to represent the interests of the Natives. The three Alaskans were there to protect the interests of the state. The conservationists relied upon Representative Udall primarily, but he got assistance from Representative Saylor. The oil industry's strongest defenders on the committee were the Alaskans and Representatives Kyl and Steiger, although the Alaskans put the interests of their state above those of the oil companies.

Meanwhile, in Alaska, Natives and non-Natives eyed each other anxiously. "Natives are becoming more and more difficult to work with and will become more so with any settlement regardless of what it is," one constituent complained to a member of the Congressional delegation. "Most of them are going to be disappointed anyway as they expect large cash awards in hand. . . ."

"I find the native land claims a gross injustice to the non-Native population of Alaska and totally racist," wrote another. "I thought the latter was something we were trying to get away from."

An organization called "The Other 80%" ran ads in local newspapers and sent letters and telegrams to Washington. "The Other 80%" opposed any settlement and hoped to force the Natives to go to court, where, they felt, the Natives were sure to get less than they were about to get from Congress.

Alaskans generally misunderstood the Congressional conference committee process because their legislature had a "free conference," at which a totally new piece of legislation could be written. A Congressional conference committee is bound by the terms the House and Senate have approved. It can modify provisions in one or both bills or work out a compromise between the two bills, but it cannot add wholly new or extraneous provisions. But many Alaskans still believed that a completely different bill could yet be produced. Different individuals and organizations across the state called for new hearings on the Native land claims and a completely new settlement bill.

While white Alaskans were in an uproar about the impending settlement, the Native leadership and Governor Egan tangled in the press. Wright accused the Governor of having made an agreement for the state to support Representative Aspinall's bill in conference. "If the state permits newcomers, miners or homesteaders to select lands before the Natives [a reference to the absence of a general land freeze in the House bill]," said Wright angrily, "we will defend the land like the Indians did in the Black Hills years ago." [122]

What had happened was that the Governor had met on November 12 with the AFN leadership to discuss the upcoming conference. At that meeting, the Governor apparently told the Natives that he favored the land selection provision in the House bill, Representative Aspinall's "some now, more later." The

Natives, who had always objected to this but refrained from saying so publicly to preserve the facade of unanimous support for the bill, were furious. They wanted first choice on all 40 million acres, they told the Governor. And they threatened that, if the Aspinall land selection provision prevailed, Native groups and villages would fight the state's land selections with suits that could cloud all titles in Alaska for twelve years. Furthermore, an AFN memorandum circulated about this time warned that, unless the Natives got first choice on 40 million acres, they would combine forces with the conservationists to make sure the state's 1983 deadline for completing its selections was not extended. This was a formidable threat since, with lawsuits pending, it would be hard for the state to get patent to any land picked during the next twelve years, leaving them in 1983 with about 26 million acres or one-fifth of what the state had been promised in the Statehood Act of 1959. Even if this did not happen, bitterness could tear apart the social fabric of Alaska.

Throughout this, the special interests were also at work. The miners wanted to make sure that prospectors could continue to locate metalliferous metals on withdrawn federal land, as they had throughout the freeze. Alyeska, which was worried about the exposure a transportation corridor would give its project, wanted to make sure that there was no implication in the final bill that the pipeline company would have to comply with any of a land use planning commission's procedures before its project could begin. Foster and another Alyeska lawyer, Quinn O'Connell, suggested specific language which would protect the oil companies' interests.

On November 16, the day before the conference met for the first time, Wright sent a letter to all its members. He made land the big issue. The Natives wanted thirty million acres of village land and ten million for the corporations, the Gravel-McGovern amendment which was never offered, Wright said. They wanted to pick all their land first. Any delay, Wright said, might lead to "state-Native land conflicts." In addition, the Natives wanted twelve "strong" regional corporations. "These corporations should have broad business and investment powers," Wright wrote. "They should not be structured to produce an excessive pass-through or dissipation of funds. The Alaska Natives intend to conserve the proceeds of the settlement fund, maximize the return on these proceeds, and use the returns to finance necessary projects. The fund should not be used for a per capita dole. Rather economic leverage should be obtained through proper investment practices at the regional and statewide level." [123] What Wright had in mind were strong, independent economic entities that could wheel and deal, just like any other business. The Native leadership did not want the bulk of the benefits to go to the villages because they feared the money would be dissipated there. They had, by this time, resigned themselves to the fact that most of the land would be tied to the villages, as the state wished, but they wanted the proceeds of its development to go to the regional organizations.

Essentially, the land selection process and the distribution of finances were the points of conflict with the state. Senator Stevens and Representative Begich generally supported the state's position. Senator Gravel did not. Senator Gravel had promised to fight for the 30–10 formula in conference. Now, on the eve of the conference, Senator McGovern, who was not a conferee, sent a letter to those who were. "The amendment was withdrawn on the floor on the basis of strong assurances from the committee chairman that the Senate conferees would fight for that position in conference," Senator McGovern wrote his colleagues, "and that under the conference rules it could be achieved. I look forward to the Natives' position being adopted by conference committee. . . ." [124]

In the meantime, at Representative Aspinall's request, Senator Stevens had written a memorandum outlining his ideas on how to reconcile the two bills. He suggested taking the House's Natives–state–Natives land selection process, doubling the amount of land for the villages from 18 to 36 million, and allowing the Natives to take any land in the state, including tentatively approved state land, provided it was contiguous to the villages. The regional corporations would then get four million free-floating acres, to be picked roughly in accord with the Jackson formula: one million acres of timber, one million acres of mineral land, one million acres of recreational land, and one million acres to straighten out individual inequities.

Representative Begich, as he had throughout the negotiations, remained noncommittal. Publicly, he supported the use of Pet Four to give the Natives some economically valuable land. This was popular with both the Chambers of Commerce and the Natives, he said, and allowed him to win the support of both sides. Actually, he knew that Representative Aspinall would never accept that use of Pet Four. In fact, almost everyone at the conference understood that.

On the eve of the first meeting, Representative Aspinall seemed to be having second thoughts about the speed with which the bill should be enacted. At a Democratic Study Group reception the night of November 16, Representatives Meeds and Begich told AFN lawyer Weinberg that the Chairman proposed to delay the conference until after Christmas. But oil lobbyists quickly changed his mind. On the following morning, the conference started work as planned.

First, the conferees elected Representative Aspinall chairman of the conference, a wise move because of his vanity and because, by this time, Senator Jackson was seriously considering a try for the Presidency and was unable to attend most of the meetings. The conferees also decided in what order to take up the various aspects of the settlement. Land was to be first, then came corporate organization, administration, land use planning, money, and other issues in that order. The conferees agreed to try to resolve all their differences by December 4, a little more than two weeks away, so that the bill could be passed that year.

But the Thanksgiving holidays intervened and the committee did not meet again until Tuesday, November 30. As planned, they discussed land. Basically, the conferees considered two proposals. One was put forward by Representative Aspinall and gave the

Natives first choice of 23 million acres, to be picked from a withdrawal area of nine townships around every village. Then the state would pick 30 million acres. However, this included the land it had already selected and that which had been patented or approved for patent, so the state actually would get only a few million acres it had not already chosen. After the state had made this selection, the Natives were to get 17 million more from a 25-township withdrawal area and, at the same time, the state would complete its selections elsewhere.

The other proposal was made by Senator Bible, who acted as leader of the Senate conferees in Senator Jackson's absence. Senator Bible suggested that the Natives be given 26 million acres on the first go-round. The state would then get 28 million, including previously selected lands, and after that the Natives could have another 14 million. In both cases, the land was closely tied to the villages. The conferees were not even discussing the AFN's 30–10 formula. In fact, it was never seriously considered, despite Senator Gravel's promise.

The day after this meeting, Wright wrote another letter to the conferees, in which he stressed the importance of allocating the land on the basis of the amount of land claimed, rather than by population. This reflected the concern of the Eskimos on the North Slope, who were few in number but claimed a lot of land. Wright wanted the money allocated the same way. Furthermore, the ASNA, whose oil-rich land would support the state's portion of the financial settlement, wanted 50 percent of the proceeds from land in their region returned to them rather than distributed among the Natives at large. Not to distribute the money and land this way would endanger "the delicate compromise achieved within AFN," Wright warned.[125]

On Thursday, December 2, the conference met again. This time the conferees reached a tentative agreement on the first round of land selections. The Natives would get 25 million acres, to be picked from a withdrawal area that covered twenty-five townships around every village, they decided. Then the state could pick 36 million acres—the 26 million it had already selected plus 10 million more. The figures changed somewhat later, but the conferees obviously had reached a basic agreement on the land selection process. One element of it was that the state's previous selections were protected, thus giving the state first crack at Alaska's 375 million acres while appearing to give the Natives first choice.

Since the next item on the agenda was the organization and administration of the settlement, Senator Bible told Senator Stevens that the state had better agree in advance on a position so that the Alaskan delegation would not take up the conference's time arguing among themselves. As a result, the three members of the delegation met in Senator Stevens' Capitol Hill office on Saturday, December 4 with Governor Egan and Attorney General Havelock. The Natives were not invited. The meeting went on all morning and, toward the end, the men called in Senator Stevens' secretary, Celia Niemi, and dictated the agreement they had reached. Miss Niemi transcribed her notes after the meeting and, on Monday morning, typewritten copies of the agreement were given each member of the delegation. The Governor, who had returned to Alaska, received his copy by Telex. The plan was that Representative Begich would show his copy to Chairman Aspinall and Senator Stevens would show his to Senator Bible. Once the two men had seen the agreement, it would be shown to the AFN, according to Senator Stevens. However, before Representative Begich could talk with Chairman Aspinall, Senator Gravel gave a copy to the AFN leaders, who were furious.

From the Natives' point of view, the most devastating part of the agreement was the way in which the state wanted to distribute the proceeds of the settlement. Ever fearful of setting up economically, and thus politically, powerful regional organizations, the Congressional delegation and the Governor had decided that 75 percent of the "capital" (federal payments and state royalty payments) received by the regional corporations should be passed on to the villages. The corporations would thus be able to invest only 25 percent of it, or about $250 million. The Natives wanted the arrangement to be just the opposite. Furthermore, although the income from investments and the use of the land was to go to the regional corporations, the state wanted only 20 percent of it reinvested by them. The rest would go on to the villages. Also, any income derived from the sale of land or the development of it in a given region was to be distributed throughout the state, with only 20 percent remaining in the region of origin, which was exactly what the North Slope Natives did not want to see happen.

As Attorney General Havelock said later, everyone at the meeting was aware that the AFN position on the distribution of "capital" grew out of the established Native power structure and was a delicate balance of the interests of the various "regional dukes," as he called them. This was partly selfish, partly altruistic as far as the Native leadership was concerned. The power structure that would be perpetuated under the AFN settlement would be the regional power structure within it, so current leaders had a vested interest in its preservation. At the same time, they also saw it as the way to give the Native people the greatest economic and political leverage in state affairs. They feared that giving most of the economic and political power to the villages would merely dissipate the benefits of the settlement. Havelock said that all five men at the Saturday meeting agreed that, under the AFN settlement, the villages would get the short end of the stick. He said they were worried that some areas, like the North Slope, would prosper at the expense of others.

The Native leadership denounced the secret meeting. They said they had been betrayed. The state agreement had cut the heart out of the AFN plan and indeed endangered the political balance within the organization, which was held together by a delicate compromise on the matter of distribution of capital. At the Natives' insistence, Senator Gravel went back to the conference and got the percentages modified a little.

The agreement also covered a variety of other administrative issues. There would be twelve regional corporations which could merge into no less than seven if they wished. The final decision on the disposition of state lands was to be made by the Governor, in spite of the land use planning commission. Prospecting and mining were to continue as they had before. The agreement allocated $1.7 million to pay attorneys' fees and retained the Senate bill's $600,000 to repay loans to the AFN.

Looking back at the meeting, the men who were present do not agree about its importance. Havelock felt that, once the meeting was over, the bill had been written except for the land use planning sections. He saw the chief purpose of the meeting as being to prod Senator Gravel into line with the rest of the delegation and he said that Senator Stevens and Representative Begich did most of the prodding. However, Representative Begich thought the meeting was little more than an exchange of views and that the most important thing was that he told the other men how to handle Representative Aspinall. Senator Stevens thought that, while the agreement's value as a working paper was undermined by its disclosure, it remained the final statement of Alaska's position and was regarded as such by Senator Bible and Representative Aspinall.

The conferees met Monday and Tuesday, and by Wednesday, December 8, had reached an agreement on the land and the money and how both should be distributed. They decided that the Natives should get full title to 40 million acres, to be selected over a four-year period. First, the Native villages would get 22 million acres, to be allocated on the basis of population. The village lands would have to be compact and contiguous and were to be chosen from a 25-township withdrawal area around each village. Villages in wildlife refuges and Pet Four would be limited to a 9-township withdrawal area, would only be able to get a maximum of 69,120 acres, and would get no mineral rights, the latter being a concession to Representative Hébert. Instead, they would get "in lieu" mineral selections from the public lands nearest the villages. During a three-year period, the state would not be able to select lands within the 25-township withdrawal area. On the other hand, no village would be allowed to select more than three townships of land previously picked by the state.

At the end of the three-year period, twelve regional corporations would be allowed to pick 16 million acres on a land loss basis, so that the total of the village selections and the regional corporations' would be in the same proportion to 40 million acres as the land that region claimed had been to the total claimed by all the Natives. This last 16 million acres, to be picked before the end of the fourth year, would have to be taken from the same 25-township area. To allow state selections to go on simultaneously, the conferees set up a checkerboard pattern and let the Natives and the state pick alternating townships during the fourth year. The Natives were also to get an additional two million acres for "hardship" cases, bringing the total to 40 million.

What the conferees had done was to take the Aspinall plan for a two-tiered selection and then limit the area from which the Natives could make selections to land around the villages, as Senator Stevens had suggested. This gave the Natives first crack at their entire 40 million acres, but only within a limited area. At the same time, the state got almost all the 26 million acres it had already chosen. You could say the Natives got first choice or you could say the state got first choice, depending upon whom you were talking to, a convenient sort of arrangement for politicians to have made. Outside the 25-township withdrawals, the state could go ahead with its land selection program unhampered. For the most part, this was where the land desired by the state lay. Gone was any idea of Native land for economic potential, whether the free-floating selections the Natives had sought or the seven and a half million acres of timber, mineral land, and recreation resources Senator Jackson had proposed.

While the Natives got what they said they wanted—first crack at 40 million acres and full title to their land, the state also got a good deal—the land it had already picked, plus no land freeze and the right to finish selecting its land immediately, for the most part.

The state lost little choice land by the settlement. And it was what the state had wanted—a village settlement. The bill created almost two hundred small enclaves wherever there were villages, leaving land management decisions largely up to the villages, although the subsurface rights to their lands went to the regional corporations.

The conferees decided to give the Natives $462 million in federal funds over eleven years, a compromise between the Senate and House figures and schedules, and $500 million from a two percent royalty. All funds would go to the regional corporations, which would keep 50 percent and pass 50 percent on to the villages in their regions. Income from Native lands would be distributed in such a way that 30 percent remained in the region of origin and the rest was divided among the other regions. This was the result of the Natives' outrage at the state's secret agreement on how the capital should be handled.

There were to be twelve regional corporations and no statewide corporation. The conferees also gave the Natives the option of creating a thirteenth corporation for nonresident Natives. The villages were to incorporate as well.

What remained to be determined were the federal government's rights and the point at which it could step in and claim them. The conservation groups kept up unrelenting pressure on the conferees and their case was ably argued within the conference by Representative Udall. Many conferees still feared the bill might be defeated in Congress if it did not contain some sort of combination of the Bible and Kyl amendments. The Kyl amendment, in effect, was a continuation of the freeze. The Bible amendment required the Secretary of the Interior to make periodic reports about what he wanted done with Alaskan land, so that Congress might act. Representative Udall was adamant that the federal government should get a chance to study the state for possible new parks and refuges before the public lands were opened to development. The state wanted no freeze at all. As a condition for Congress' lifting the freeze, the state representatives agreed that some land could be withdrawn for possible inclusion in the national park or refuge systems. Senator Stevens said later that everyone had a fairly good idea what land Representative Udall was talking about and that, for the most part, it was land already withdrawn for special purposes or classified for inclusion in park lands, like the proposed Gates of the Arctic park in the Brooks Range. These areas added up to roughly 50 million acres. On December 9, the conferees decided to give the Interior Department 80 million acres, in order to give it some leeway.

Representative Udall protested that 80 million was not enough. He said that, in addition to those areas already classified for study, there were eight or ten other parts of the state that should be protected. But Representative Aspinall, who had already announced that if the conference failed to reach an agreement that day it would not meet again until after the first of the year, warned Representative Udall that he did not want Alaska "blanketed" by the federal government just to keep the state from finishing its selections. So the conferees gave the Secretary authority to withdraw up to 80 million acres for possible inclusion in one of the conservation systems. First, this land had to be withdrawn within ninety days after the bill became law (there would be a total land freeze until then). Six months after that, the Secretary would have to make a further decision about which lands to keep for parks and ranges. He could keep all 80 million if he wished. Furthermore, the conferees specified that this withdrawal should include all land already classified for consideration, like the Gates of the Arctic. But the conferees also gave the Secretary the power to withdraw other public lands ninety days after the bill became law if necessary "to insure that the public interest in these lands is properly protected." The purpose of this part, Section 17 (d) (1), was to prevent a land rush, which certain conferees and members of the Committees' staffs feared.

The bill was hastily drafted so Congress could pass it before the Christmas recess, and, as the state discovered soon after its passage, Section 17 (d) (1) could be construed to mean that the Secretary had the right to withdraw the land in the public interest before the state completed its selections. The language was ambiguous. The only mention of state selection rights is in a section which specifies that these "public interest" withdrawals will not affect Native and state selections within the 25-township withdrawal areas. There is no mention of land elsewhere in the state. It boiled down to a question of legislative intent: Did the conferees mean to give the federal government the right to halt state selections?

The statement by the conference committee is ambiguous, too.

> It is, however, a very broad and important delegation of discretion and authority and the conference committee anticipates that the Secretary will use this authority to insure that the purposes of this act and the land claims settlement are achieved, that the larger public interest in the public lands of Alaska is protected, and that the immediate and unrestricted operation of all public land laws 90 days after date of enactment—absent affirmative action by the Secretary under his existing authority—does not result in a land rush, in massive filings under the Mineral Leasing Act, and in competing and conflicting entries and mineral locations.[126]

Although it was then unclear whether the state was taken care of, the miners were, as Senator Stevens had promised. All the public interest withdrawals, except those from which the Natives were to choose land, allowed miners to continue to locate minerals. The final bill also provided that any mining claim within those withdrawals which had been initiated before August 31, 1971, was protected. Further, it directed the Secretary to promptly issue patents for homesteads, trade and manufacturing sites, and small tract sites to anyone who had fulfilled the requirements but been unable to get a patent because of the land freeze. And it provided that anyone who had made a lawful entry onto public lands before August 31, 1971 could have his interests protected until he had had time to perfect his claim.

Alyeska was also taken care of. The final bill contained

Senator Gravel's land use planning commission but not Senator Jackson's transportation corridor. However, in its statement, the conference committee noted that the Secretary already had the authority to withdraw the land for the corridor. Should he withdraw such a corridor, the report continued, "The state and the villages and regional corporations may not select lands from the area withdrawn for the corridor." This took care of Alyeska's two fears: that the state or the Natives would pick land along the route and further delay the pipeline and that the Secretary would have to consult the planning commission before granting a permit for the pipeline.

There were a few other odds and ends in the bill. At Representative Aspinall's insistence, any attempt to alter the federal leasing process in Alaska was dropped. The state had one year in which to bring a suit contesting the settlement in federal district court in Alaska. Should the state initiate litigation or voluntarily become a party to litigation, its land selection rights were to be suspended as long as the suit was pending. But state selections would not be held up if the Natives sued, nor could conservationists bring suit in an "unfriendly [to the state of Alaska]" court in the District of Columbia. Attorneys' fees were limited to a total of $2 million and individual lawyers would have to submit their bills to the Court of Claims first. Consultant fees were limited to $100,000 in all and $600,000 was set aside to pay the Natives' expenses, not including lawyers and consultants. The final bill included a provision that no settlement funds were to be used for "propaganda" or for "intervening in (including the publishing and distributing of statements) any political campaign on behalf of any candidate for public office," a "Hatch Act" clause insisted upon by Republican Stevens. On December 13, the conference filed its report and accompanying statement by the conferees. Only Representative Saylor refused to sign.

The following day, December 14, the House approved the settlement by 307 to 60, a margin of 247 votes. An hour later, the Senate, too, passed it. In both cases, discussion on the floor was desultory. Everything that could be said had been said. Or nearly everything, as it turned out.

Although the settlement was probably as generous as political and economic realities allowed, the Natives were not entirely happy with it, particularly the Eskimos on the North Slope. Oil had been responsible for the settlement, but the Natives had ended up with no known oil land.

On December 18, 1971, President Nixon signed the bill and the Native land claims settlement became law.

On the same day, the Alaska Federation of Natives met to formally approve the legislation. It was a big occasion and, by means of special direct lines, the President spoke to them as he signed the bill. Then the regional organizations voted on whether they approved the settlement. Only the Arctic slope Natives voted No.

11

ON WITH THE PIPELINE

As the claims settlement progressed through Congress, the Interior Department grappled with the issue of the pipeline. The settlement cleared away only one roadblock. The department still had to comply with the National Environmental Policy and Mineral Leasing Acts.

Comments on the Interior Department's January 1971 draft environmental impact statement on the project underlined a question which had been raised at public hearings in February: Shouldn't the pipeline go through Canada rather than Alaska? It was not a question the Department was anxious to explore. Throughout 1971, Solicitor Mitchell Melich insisted that the Department was required to consider only the application the oil companies had submitted to it, the application for permission to build an Alaskan pipeline. But Melich changed his thinking after the Appeals Court for the District of Columbia ruled in January 1972 that NEPA required the broad consideration of alternatives to any proposed federal action.

Clearly there were some drawbacks to the Canadian route. A pipeline up the Mackenzie River would be nearly twice as long as one through Alaska. It would go through nearly twice as much permafrost, although not necessarily the most difficult kind from an engineering point of view. It would pass through a foreign country. And its construction would delay the oil companies' profits by several years at least.

But there were also obvious advantages. A Canadian pipeline would avoid the numerous earthquake zones through which an Alaskan pipeline would pass. It would not involve tanker traffic up and down the rugged west coast of Canada, endangering both Canadian and Alaskan fisheries. Finally, there are an estimated 26 trillion cubic feet of natural gas at Prudhoe Bay. At that time the companies had decided that taking the gas by pipeline to Valdez and converting it there to liquefied natural gas for shipment south on special tankers was too expensive. (Since then El Paso Natural Gas has decided that it may be economically feasible to transport gas to Valdez and liquefy it.) The gas would have to go by pipeline from Alaska through Canada to the American middle west. This meant that there was certain to be one pipeline along the Mackenzie River. In all probability there would also be an oil pipeline from the Canadian Arctic some day, running along the same valley. Conservationists proposed that all these pipelines from the Arctic be constructed in a single corridor, thus minimizing the danger to the environment.

The Canadian route was the only alternative to which the oil companies had given real consideration, but they discarded it sometime in late 1968 or early 1969. Alyeska's predecessor, TAPS, had commissioned several studies of the Mackenzie River route, most of them brief. All of the studies concluded that a Canadian pipeline was economically and technically feasible. But the companies stuck to their original decision that a Canadian pipeline would be too long and too expensive and would involve too many different governments.

In 1969 and 1970, Standard Oil of New Jersey, now the Exxon Corporation, had had an expensive flirtation with another alternative, the Northwest Passage, a series of deep water channels which are covered with ice most of the year. The company spent $50 million to send a huge ice-breaking tanker, the *Manhattan*, on two voyages through Canada's northern islands to see if it was possible to get Alaskan oil to U.S. markets that way. After considerable fanfare, Standard Oil of New Jersey abandoned the idea in October 1970.

The Interior Department's 1971 draft environmental impact statement devoted only two pages to the so-called Canadian Alternative. It pointed out that, while specific Alaskan environmental problems might be avoided, similar ones would be encountered in Canada and that a Canadian pipeline would take from two to four years longer to build. Furthermore, an oil supply which had to travel through Canada "is not wholly within the control of the U.S." [127] "In summary," the report concluded, "there is no transportation mode that offers either less risk of damage to the environment or is as economically attractive as the [Alaskan] pipeline." [128]

From the time he took office in 1971, Secretary Morton was uncomfortable about the pipeline. Clearly he disapproved of his predecessor's open advocacy of it. The draft impact statement had been released before he took office and the Secretary more or less disowned it at his confirmation hearings. He also refused to be pinned down to any sort of timetable for approving the project although both the industry and his inherited bureaucracy were pushing for its speedy completion. Secretary Morton said he was a long way from deciding that the pipeline was the best way to get the North Slope oil to market. In practice, however, he did little to alter a system which was predisposed to grant a construction permit.

Secretary Morton's public remarks came as a blow in Alaska, where pipeline advocates were as optimistic as ever, despite the failure of a number of desperate schemes to get the project moving. In 1970, Governor Miller had tried to get the legislature to appropriate $120 million for construction of a state road to the North Slope, for which the oil companies were to reimburse the state once they got their permit for the pipeline. The legislature appropriated the funds but added a proviso that the oil companies were to repay the state within five years, regardless of whether or not they had gotten permission to build their pipeline. The same legislature had pushed through a new sliding scale severance tax, which meant that the oil companies on the North Slope would be paying about 20 percent in taxes and royalties when their wells finally started producing. While there was nothing the companies could do about the severance tax, they refused to build the road on the legislature's terms. Governor Miller considered convening a special session of the legislature to reconsider the road, then realized that he had lost, and let the road plan slip into well-deserved oblivion. Then, in January 1971, after Governor Egan's election, AFN president Don Wright proposed that the Interior Department grant the land along the pipeline route "in trust" to the Natives who claimed it. The Natives would then lease it to Alyeska until the claims were settled, thus allowing the project to begin. Neither the industry nor the state endorsed this proposal. And one of the Native groups in question, the Arctic Slope Natives Association, objected vehemently. "You want my land, people, you pay for it, if you want to develop it," said Joe Upicksoun. "My people are not concerned about a phony economy like the Caucasians, like the state of Alaska." [129]

Meanwhile, the Canadian government was growing increasingly interested in an oil pipeline up the Mackenzie River. Such a pipeline could give development in the Canadian Arctic a boost. The Canadian government was committed to development there but could not afford to build such a pipeline itself. Nor could Canadian companies. At the same time, public feeling against the Valdez-to-Vancouver tanker route was running high. Jack Davis, Minister of Fisheries and soon to become Canada's first Minister of the Environment, endorsed the Canadian alternative early in 1971 because of objections to the tanker route. There were rumors that Davis intended to speak at public hearings on the pipeline in Washington in February. In the end, Davis did not come, although David Anderson, a Member of Parliament from Vancouver who built a one-man crusade around this issue, did. However, Anderson came as a private citizen, not a representative of the Canadian government. "Your gain, our risk," he told the Americans. He was liberally wined and dined by oil company representatives in Washington but their attentions did not deflect him. At the same time, Canadian politicians started lobbying in Ottawa for the Trans Canada pipeline. While visiting Washington later that spring, Resource Minister J. J. Greene informally invited applications for permission to construct a Mackenzie River Valley pipeline from the major oil companies. Eighteen Members of Parliament wrote the Interior Department protesting the tanker route. Demonstrators gathered at the U.S. customs office on the border in British Columbia to protest the Alaskan pipeline. In late February, External Affairs Minister Mitchell Sharp wrote Secretary of State William P. Rogers asking that Canada be consulted before any final decision was made on the Trans Alaska Pipeline.

Alaskans generally viewed the interjection of the Canadian alternative as a red herring. "The 'preservationists' opposing the pipeline first zeroed in on the permafrost and threat to wildlife considerations," commented the pro-pipeline *Fairbanks News-Miner* after the public hearings. "They have since expanded their objections to include dire threats of what will happen to the fisheries in Valdez and oil spills off British Columbia. They undoubtedly will find other objections when the old ones have been answered." [130]

Alaskans also feared Canadian competition for the oil industry's favors. Governor Egan observed tersely that Canadian "maneuvering" was to be expected. "I believe some Canadian officials are flatly opposed to meaningful Alaska development. Period," he said.[131]

Early in 1971, the true dimensions of the state's fiscal problems had begun to emerge. In spite of the $900 million infusion into the budget from the great lease sale in 1969, the state's general fund was in serious trouble. Alaska had gone on a spending binge. Its budget was larger than its intake in taxes. True, the state was spending money on basic services that residents of most states take for granted as well as unnecessary public works projects, but, since statehood, Alaska's budgets had increased at an average of 21 percent a year. Between fiscal 1970 and 1971, the budget soared 77 percent. About 28 percent of the $300 million plus budget for fiscal 1971 came from its $900 million nest egg, thus decreasing the capital on which the state earned interest. Assuming that future budgets rose 15 percent each year, the general fund would contain only $489,000 by June 30, 1976 and the state would start operating on a deficit, according to Governor Egan's calculations.[132]

The unknown element was how soon the state would start receiving revenue from North Slope oil production, something which was impossible until the pipeline or a substitute for it had been completed. The general assumption was that the money from severance taxes and royalties would start coming in during fiscal 1975–76, if pipeline construction began by 1972. If the money did not start coming in, the state's treasury would be in a sorry shape. A delay beyond 1977 in the pipeline's completion would wreak havoc with Alaska's finances. Therefore, the state's objection to the Canadian pipeline was that it would take at least two years longer to build than the Alaskan one, thus delaying the time when revenue would start flowing into the state treasury from the North Slope. Alaska needed that money as soon as possible.

Secretary Morton made sure the new impact statement his Department was drafting would include a pro forma consideration of alternatives. He ordered a study of the impact of tanker traffic on Canada's coast and indicated that he thought the Department should consider the Mackenzie Valley route too, regardless of what the Solicitor advised. On March 12, the U.S. Environmental Protection Agency, commenting upon the draft statement, urged that the Canadian alternative be studied more closely.

Secretary Morton was not opposed to the pipeline *per se,* he said. "I'm not particularly for the Canadian proposal or against it," he told one Congressional Committee. "I just think we ought to take a look at it." [133]

In late March, at his insistence, five of the North Slope oil companies—ARCO, Humble, Mobil, BP, and SOHIO—sent high level representatives to Ottawa to talk with Canadian officials about a Canadian pipeline. It was a rather one-sided discussion since the oilmen reportedly did most of the talking. When they returned to Washington, they reported directly to the White House, where Peter M. Flanigan, the presidential assistant in charge of matters related to the oil industry, was very interested in the Alaskan pipeline. Four company presidents plus SOHIO's chairman of the board and Alyeska president Edward Patton explained once again the difficulties the companies had with the Canadian route. They were the same as they had been in 1969: too long, too expensive, and too many governments.

At this point, April 1971, the oilmen were still hoping to get a construction permit for the Alaskan pipeline by late summer. They hoped the Native land claims and the conservationists' suit would be resolved in time for construction to start in 1972. "You ought to quit worrying about Canada," Patton told the Alaska legislature's Joint Legislative Pipeline Impact Committee in Juneau on April 2, about a week after the oil companies' visit to Ottawa.[134]

In the meantime, another foreign country had entered the pipeline picture—Japan. Actually, Japan had been on the Alaskan economic scene for years. In 1970, Japan was Alaska's biggest customer for exports, purchasing more than $100 million worth of pulp, lumber, fish and fish products, natural gas, and ammonia.[135] Canada, which was the state's second largest customer, had bought only $10.4 million worth of products. And there were plenty of Japanese imports in Alaska, notably the rusting piles of 48-inch pipe at Valdez, Fairbanks, and Prudhoe Bay. The Japanese also sold Toyotas and Datsuns in Alaska. A Japanese company owned a pulp mill in Sitka. Japan Air Lines flew in and out of Anchorage frequently. Japanese companies were joint lease holders on the North Slope.

But Japan was also a potential customer for North Slope oil. Japan is the world's largest importer of oil. Over 85 percent of its petroleum comes from the Middle East. If, as seemed possible to some economists, the U.S. West Coast could not use all two million barrels of Alaskan oil that would arrive daily when the pipeline was in full operation, one possibility was to sell the surplus to Japan. However, the oil companies and the Interior Department said little about this because it made a mockery of their argument that the pipeline had to be built quickly to supply the West Coast with oil.

All spring, Secretary Morton was under tremendous pressure from the industry to grant the construction permit and from conservationists not to grant it. The Secretary fluctuated wildly in his public utterances about the project. "I don't favor the damn thing much at all," he said in Seattle in June. "I don't think you need all that oil out here. I think it's an awful lot of money to spend for 30 or 40 million barrels of oil." [136] At Senator Stevens' insistence, he quickly revised this remark.

Later in June, the Secretary flew to Alaska to have a first look at his domain. His visit started at Fairbanks, where he was greeted at the airport by a delegation of residents wearing "Let's Build It Now" buttons. Then he toured the pipeline route and the oil fields. From Prudhoe Bay he went to Barrow, where he told Native leaders that he thought the federal government probably should not have allowed the state to select any land until the claims had been settled. The Barrow Eskimos, then threatening a suit over who really owned Prudhoe Bay, cheered. As the Secretary flew south to Anchorage, the Interior Department in Washington announced that it was extending the land freeze another six months or until Congress passed a claims bill, whichever came first. In Anchorage, the Secretary talked with all the right people: the Republican Women of Anchorage, the city and state Chambers of Commerce, the Rotary Club, the Press Club, and the Petroleum Wives Club. He had dinner with his predecessor, Walter Hickel, and breakfast with the Governor. He flew over the Wrangell Mountains, met with the Cordova Fishermen's Union, which had recently joined in the conservationists' suit to halt the pipeline (Cordova is a coastal city near Valdez), and toured Valdez, Bethel, Dillingham, the city of Kodiak, and the Kenai Moose Range, a million acres of which the federal government then hoped to turn into a wilderness area. He returned to Washington saying he now understood more about Alaska, a remark interpreted by Alaskans to mean he agreed with them about the pipeline. He also said he had a "sense of confidence" in the companies' desire not to damage the environment. And he predicted that the final environmental impact statement on the project would be completed in mid-September. But the Department did not get all the material it needed from Alyeska until late September (and even then there was some question about whether the company had really answered the Department's technical questions). The deadline for release of the completed impact statement was pushed back to Christmastime.

In the meantime, attorneys for the federal government sought to have the conservationists' suit moved from Washington, D.C. to a friendlier court in Alaska. It was a little late for that, Judge Hart noted in late July, as he denied their request for a change of venue. However, he allowed both the state of Alaska and the Alyeska Pipeline Service Company to enter the case on the side of the defendant. In addition to the Cordova fishermen, David Anderson and the Canadian Wildlife Federation had joined the three environmental groups as plaintiffs in the suit.

After worrying about the state's impending bankruptcy, Governor Egan told the parent companies of Alyeska that summer that the state was thinking of owning the pipeline. The Governor's idea, which had evolved from a little-noticed study commissioned by his predecessor, called for financing the pipeline with a $1.5 billion bond issue and contracting with Alyeska to do the work. The Governor's aides had reassessed the 1970 report, as well as the advice of another of former Governor Miller's consultants, Walter Levy, who talked about varying degrees of state control over and equity in the pipeline but not outright ownership, and determined that there were additional advantages to state ownership. Owning the pipeline would allow the state to better assess and control the wellhead cost of North Slope oil, on which the state's royalty was determined. Although Governor Egan's proposal shocked the oil companies and some Alaskans, he had hinted at it in the past. In 1970, while campaigning for Governor, he had told a reporter, "We must really, in a serious way, explore the advantages of state ownership of a pipeline to Valdez."

Now the Governor calculated that owning the pipeline could mean as much as $100 million more a year in revenues to the state. This was the first tentative step in a series the Egan administration was to make toward regulating the pipeline as no existing oil pipeline had previously been regulated. It marked a growing disenchantment with the mystique of Oil.

Almost as soon as the claims settlement was signed by the President in December 1971, lawyers for conservation societies, the Interior Department, and the state of Alaska realized that, while it was unclear, the new law apparently gave the federal government authority to halt all future state land selections on March 18, 1972, ninety days after the settlement bill became law. At issue was the meaning of Section 17 (d) (1). That hastily written addition to the land claims bill which allowed the Secretary of the Interior to set aside land for parks and other public interest uses was extremely ambiguous.

The Alaskan Congressional delegation thought they had agreed that the federal government should get 80 million acres of land, most of which had already been classified for possible inclusion in one of the existing national conservation systems. But they did not think they had agreed the federal government should choose its land before the state got its. They thought only Native village selections were to precede the state's and then only in prescribed areas. But the men who drafted Section 17 (d) forgot to spell out selection priorities.

On January 11, 1972, not yet a month after the claims issue had been settled, conservation-minded Representatives Saylor and Udall wrote Secretary Morton an eight-page letter, in which they laid out their interpretation of that section. "The opportunities implicit in these provisions go far beyond any single opportunity of this kind ever before given to the Department of the Interior," the Congressmen wrote. "We find the prospects immensely exciting, and will want to be fully supportive and encouraging to you in every way to see that they are fully realized." [137]

What they proposed was this: The Secretary should make the broadest possible "public interest" withdrawals before March 18 "to assert in a positive way the interest of the American people in sound federal land use and land disposal decisions in Alaska." [138] Section 17 (d) (1), they said, was a new classification authority which allowed Secretary Morton to, in effect, reinstitute the old Udall land freeze in Alaska and then lift it at his will and at whatever pace he desired. These withdrawals would give Secretary Morton the power to "assist" the state in making its selections.

The state had anticipated the Udall-Saylor letter. A few days later, on January 21, and again on January 24, the Governor filed applications for 77 million acres with the BLM, thus completing the statehood land selections. The speed with which the state acted surprised no one. The Division of Lands had been at work ever since 1966 determining which land to pick. It was merely a matter of completing and submitting the applications. The state was perfectly frank about what it was doing. Attorney General Havelock said, "These provisions did force the state to accelerate its selection process, just as they have forced the federal agencies involved to accelerate the process of final classification." [139]

Most of the newly selected land was either mineral land or land with other commercial value. The Governor's actions enraged conservationists since the land the state chose included almost the entire proposed Wrangell Mountains National Park, much of the proposed Gates of the Arctic National Park, and proposed additions to the Mount McKinley National Park. In February, the Sierra Club, the Wilderness Society, Trout Unlimited, and the National Audubon Society took out full-page advertisements in the nation's leading newspapers. ALASKA UP FOR GRABS read the caption, and beneath it were drawings of Secretary Morton, Teddy Roosevelt, an Alaskan brown bear, and a bull moose. "These are national treasures, these wildlands of Alaska," said the ad. "One man alone can protect them now. But unless Rogers C. B. Morton acts now—before March 18—much of them could be lost to all of us, for all time to come.

"A giant land grab, the likes of which the world has never known, is now underway. . . . On March 18, the gun goes off and the Alaskan land rush begins. The opportunists and private interests are poised on their haunches; and, surprisingly, the state government of Alaska has already jumped the gun. But first claim on these treasured lands belongs, in fact, to all the American people, to be held in sacred trust and preserved as national wonders for all time to come." The advertisement ended with a coupon the reader could clip and mail to the Secretary if he could not compose his own letter.

On February 22, the officers of the Environmental Defense Fund had a letter hand-delivered to Secretary Morton in which they laid out a case for the broadest possible "public interest" withdrawals. And they threatened a lawsuit, the conservationists' prime lobbying tool. "Any other action—for example, to open up any portions of such land to mining or logging—will require compliance with NEPA, including a full technology assessment and a Section 102 (c) (2) 'detailed statement' on the proposed action. . . . But extending the blanket withdrawal would not require an impact statement since such option temporarily continues the status quo and protects the environment from damage." [140] The timing of the letter was crucial. The conservationists knew that Secretary Morton was to be briefed by his staff on what to do about the Section 17 public interest withdrawals the following day. The final decision would be made about March 2.

Meanwhile, legislation to clarify Section 17 (d) and correct errors in the Land Claims Act was pending before Congress. (There were a number of errors, including one which gave the Native regional corporations rights to oil and minerals in Pet Four. The AFN did not want this error corrected.)

However, on March 15, before Congress had acted on any of

these bills, Secretary Morton withdrew 80 million acres to be considered for inclusion in the national conservation systems and another 45 million acres of public interest lands, including many forests and mineral deposits. He also set aside an additional 44 million acres for the Natives to pick land from, bringing the total Native withdrawals to 99 million acres, 40 million of which would eventually be theirs. In so doing, Secretary Morton closed to entry 44 million of the 77 million acres the state had filed for in January. A lawsuit was inevitable and the state brought one a few days later, although it was always understood that both parties would seek an out-of-court settlement through the federal-state planning commission, whose new federal cochairman was the Secretary's former assistant, Jack O. Horton.

The disputed lands were located all over the state. Some conflicts could have been predicted in advance. Included in the 80 million acres were lands around Mount McKinley National Park, land on the southern slope of the Brooks Range, land along the coast of the Gulf of Alaska, and portions of the Copper River Valley. Included in the public interest lands were portions of the Wrangell Mountains, mineral deposits on the Seward Peninsula, and land on the Alaska Peninsula. Included in the Natives' new 44 million acres were the Forty Mile area on the Canadian border and other lands west of the Copper River.

"You have met the measure of this challenge," conservation groups telegraphed Secretary Morton.[141] No wonder they were happy. The March 1972 federal withdrawals were larger than those contemplated by Representatives Udall and Saylor when they unsuccessfully proposed their land use planning amendment the previous October. The freeze had been continued temporarily in most of Alaska.

The oil industry was an interested observer at all these proceedings, since the Secretary also set aside a second transportation corridor (the route of the proposed Trans Alaska Pipeline had already been classified as one). This second corridor, some 1.2 million acres in all, ran from Prudhoe Bay eastward around the southern edge of the Arctic Wildlife Range to the Canadian border. Its purpose was to accommodate a gas pipeline from Prudhoe Bay to Chicago. It could also be used for the Alaskan segment of an oil pipeline to the middle west.

Five days later, on March 20, the Interior Department released its voluminous environmental impact statement on the Trans Alaska Pipeline. Although final decision was put off for forty-five days, the Secretary had already hinted at what it would be. During Congressional hearings on NEPA's effectiveness in early March, Secretary Morton spoke as though the pipeline was an accomplished fact. The impact statement consisted of nine volumes—six dealing with the environment, which were bound in institutional green, and three dealing with the economic and security aspects and covered with pale yellow.

The six green volumes were packed with details of possible environmental degradation, which seemed to point to the conclusion that a Canadian pipeline would be preferable to an Alaskan one from an environmental point of view. Such a pipeline would avoid earthquake zones and the hazardous tanker route. Secretary Morton ultimately dismissed both. The pipeline could be built to withstand earthquakes, he told a reporter during the 45-day waiting period, and oil pollution of the oceans was still less dangerous than DDT.[142]

The study stopped short of advocacy, leaving the reader and the Secretary, who did not read it all, to draw their own conclusions. However, the question in the pipeline suit pending before Judge Hart was not whether the Secretary had made the right decision but whether he had considered the alternatives. The report did not discuss the inevitable gas pipeline through Canada, although buried in it at irregular intervals were references to an overland gas pipeline from Prudhoe Bay to Chicago. However, nowhere was it evaluated as part of the total Alaskan transportation system. Conservationists argued that it should have been.

As indicated by subsequent depositions taken by the conservation groups who were suing to prevent the pipeline's construction, there were members of the task force which worked on the impact statement who had wanted to evaluate this total system. One of them was Dr. David Brew, chairman of the task force. In February 1972, following an Appeals Court ruling that the Department must consider alternatives that were beyond its aegis, that task force was directed to attempt some evaluation of the gas transportation system. Dr. Brew wanted to evaluate the benefits of a common oil and gas pipeline corridor through Canada as compared to the benefits of an Alaskan oil pipeline, a tanker route, and a Canadian gas pipeline. But he warned his superiors that such an evaluation would probably show the Canadian common corridor to be superior. He was not allowed to make it. There was not enough time, Dr. Brew was told.

After the impact statement was released, the Secretary had Deputy Undersecretary Horton prepare a memorandum which compared the two total transportation systems for not only the oil but the natural gas found on the North Slope as well. Horton's memorandum, called "An Alternative to TAPS" and conveniently undated, was a strong case for the Canadian route. Written for the Secretary only, it came to light in June during Congressional hearings, where it was presented as proof that the Department had considered the Canadian common corridor. It served the same purpose later in court.

The important volumes of the impact statement were the three yellow ones—the economic and security analyses. In them, the department argued that the Alaskan oil was needed on the U.S. West Coast and would go there. However, in submissions to the Department, Alyeska had indicated that, when the pipeline reached its full capacity, 1.5 million barrels of oil would go to the West Coast and the other 500,000 to "Panama." "Panama" was a catch-all term for markets outside the West Coast. In fact, "Panama" was several places. One possibility was to ship the oil by tanker to Central America, pump it through a nonexistent pipeline, perhaps in Panama, perhaps in Costa Rica, and then move it by tanker again to a refinery in the Virgin Islands for refining

(Amerada-Hess had a refinery there already). The refined products would then be sold on the U.S. East Coast at high prices. A second, considerably less complicated possibility was Japan. In fact, Alyeska president Patton admitted that, by 1980, about 100,000 barrels of "Panama" oil might well be sold to Japan. Actually, the marketing of the oil was up to the individual North Slope companies, not Alyeska. In 1970, BP had signed an agreement with a group of Japanese oil companies which would include marketing an undisclosed amount of Alaskan crude there. And Phillips Petroleum had proposed an import-for-export plan, under which the companies would trade their excess Alaskan crude for Japanese rights to Middle Eastern oil, which could then be sold for a profit on the U.S. East Coast. While there were references to "Panama" in the impact statement, its authors nowhere dealt directly with it. The truth is that the oil companies on the North Slope wanted a quick return on their Alaskan investments, now approaching $1 billion, including bonuses and exploration costs. To them, the Alaskan pipeline meant more cash sooner. They also expected the Japanese to be willing to buy oil at any price, and remarks by Japanese officials during 1971 and 1972 indicate that the companies were probably correct in thinking this.

Another consideration was the Jones Act, legislation which forbids the use of foreign-flag or foreign-built vessels in interstate commerce. Because of the union wage scale, ships built in the United States are more expensive than foreign-built ships. However, the oil companies could use foreign ships to move oil to Japan and even through Central America and the Virgin Islands to the United States East Coast (because of peculiarities in the Jones Act, refined products may move from the Virgin Islands to the mainland United States in foreign ships). Thus, "Panama" meant higher profits for the oil companies.

Earlier, one oil company had tried an end run around the Jones Act. In 1970, Union Oil of California chartered the United States–built Liberian-flag tanker, *Sansinena*, from the Barracuda Tanker Corporation. Union Oil applied to the Treasury Department for a waiver of the Jones Act so the tanker could operate between Alaska and the West Coast. The *Sansinena* had friends in the right places. Peter Flanigan, presidential assistant for oil industry affairs, had helped to organize the Barracuda Tanker Corporation while working for a New York investment banking firm and in March, 1970, the company obtained its waiver. However, the outcry which followed forced the Treasury Department to withdraw it. Union Oil returned the *Sansinena* to her owners and ordered a new *Sansinena II* from the Bethlehem Steel Corporation for use in the Alaska trade. The administration took pains to make it clear that, despite the *Sansinena* incident, it had no intention of letting foreign vessels compete with American ones in interstate commerce.

Another issue was what the pipeline would do to the Alaskan economy. Three Alaskan economists—Victor Fischer, Arlon Tussing, and George W. Rogers—found the prospects something less than rosy. In a study for the Interior Department, incorporated into the impact statement, these economists said that building the pipeline would aggravate Alaska's stubborn unemployment problem and would not lead to development of the rest of the Interior, as many Alaskans hoped.

Alaska's unemployment problem is a peculiar one. The state consistently has a higher rate of unemployment than other parts of the country, yet these figures do not indicate the full extent of the problem. Rural joblessness, or underemployment, particularly among the Native people, is not counted as unemployment in state reports. Furthermore, transients play a considerable role in the Alaska work force because of the seasonal cycle of activity in the state. It is a paradox that the greatest number of unemployed are recorded in June, because of the influx of transients and college students looking for summer work, while June is also the peak of employment.[143]

The biggest problem is in the Native villages. Politicians and oilmen had made much of the job possibilities the petroleum industry would provide for the Natives. But the oil industry is not a labor-intensive industry. Furthermore, the three Alaskan economists put their finger on a stark truth—the pipeline and the oilfields are nowhere near most of the Native villages, and most of the villagers are "unemployable" from the companies' point of view because their subsistence life style makes it difficult for them to hold full-time jobs. (One oil company, ARCO, had experimented with Native workers at Prudhoe Bay and discovered a scheme which worked fairly well for everyone involved. The company hired teams of Native workers from Barrow, flew them to Prudhoe for the regular two-week work stint, then flew them back to Barrow and picked up a new crew. The workers then had two weeks off in which to hunt and fish for their families.)

Looking at Alaska's recent past, the economists noted that "perhaps the most significant inference from the employment-unemployment relationship is that increased employment does not consistently mean fewer unemployed. In the long run particularly, the growth of employment has little effect on unemployment; the correlation, if any, is weakly positive. . . ." [144] They concluded that while there would be an increase in employment during the period of pipeline construction, there would be a sharp rise in unemployment after its completion unless secondary employment increased markedly. Whether it did or not would depend upon state spending policies.

The economists also concluded that the construction of the pipeline probably would not stimulate mining and lumbering elsewhere in the state. The most frequent argument for building a road or railroad to the North Slope had been that surface transportation would "open up" the Interior. But the economists said that the pipeline and adjacent road would not pass through significant mineral areas that were not already accessible.[145]

They decided that the real impact of the oil development would come from revenues to the state and that the degree of this impact would depend on what the state did with the money. If the money circulated in the state in the form of wages and salaries, the impact could be considerable, they said. The Interior Department had concluded that the state's profits would be slightly higher if the oil went to market through an Alaskan pipeline because the wellhead price of oil delivered to the West Coast would be slightly higher than that of oil delivered to the Chicago area, although critics like Charles J. Cicchetti and John V. Krutilla of Resources for the Future, Inc. later argued that inadequate treatment of certain factors biased the Interior Department's findings in advance. (The wellhead price is determined by subtracting the cost of transportation from the price of the delivered oil. The wellhead price of Alaskan oil delivered on the West Coast was then estimated to be between $2.29 and $2.33 and the wellhead price of oil delivered to Chicago to be between $2.20 and $2.30.)[146] The state also needed the revenues as soon as possible and the Interior Department calculated that an Alaskan pipeline could be completed two years before a Canadian one. This, according to the Interior Department, was the real argument for an Alaskan pipeline from the state's point of view.

The Alaskan economists did not try to assess whether the state would get more money from an Alaskan pipeline than a Canadian one. However, one of them, Fischer, head of the Institute of Social, Economic, and Government Research at the University of Alaska, commented, "From Alaska's standpoint, there's a great deal to be said for just clipping coupons." The state's oil revenues would be essentially the same whichever way the oil went to market, he pointed out, and, if the pipeline went through Canada, Alaska would not suffer the disruption and additional cost in state services which the Alaskan pipeline would cause. Fischer compared the pipeline to the 1964 earthquake, which gave the state's feeble economy a boost because of the influx of federal disaster aid. He wondered, wasn't it possible to get the economic benefits without the trauma of another earthquake? And he concluded that while some Alaskans, like the owners of motels, bars, and contracting businesses, would prosper from the Alaskan pipeline, others, like the villagers who came in contact with the project, would probably suffer, at least to the extent that their traditional way of life would be irreparably damaged without the establishment of a satisfactory substitute.

In summing up the pipeline's effects on the state, the Interior Department concluded that the unfavorable impacts cited by Fischer and his colleagues could be modified by state policy. The problem was that no such state policy existed. There was talk of a plan to dovetail state public works projects with the pace of pipeline construction, so that, as pipeline construction jobs were dwindling, there would be more construction work in the public sector; but it was little more than talk and even state officials were skeptical that sufficient spending could be deferred till after the pipeline was completed to have a noticeable stabilizing effect.

The impact statement concluded that North Slope oil would reduce foreign imports to the United States by an equal amount (this amount was a mere drop in the bucket of U.S. petroleum needs, although the statement did not say that); that no alternative route to market was economically more efficient than the Alaska route but that the Canadian route was equally efficient; and that the United States could not afford to wait two more years before developing the Alaskan reserves. The first and the last conclusion became the Secretary's rationale for approving the Alaskan pipeline.

Understandably, the oil companies were confident. A few days before the impact statement was released, Alyeska mailed out a handsome booklet that began: "The Trans Alaska Pipeline will pass through some of the most hostile and yet delicate country known to man. . . ." [147]

Conservationists were wary. They urged the President to ask Secretary Morton to hold more public hearings on the project. Eighty-two Congressmen made the same request of the Secretary. But he had indicated earlier that he saw no reason for more public hearings and Judge Hart had refused to order him to hold them. As he distributed copies of the impact statement to reporters on March 20, Undersecretary Pecora said further hearings would be a "circus." He added, "It would interfere with more thoughtful and rational analysis of this complex document."

Under the auspices of the Environmental Defense Fund (EDF), one of the plaintiffs in the suit to halt the pipeline, conservationists compiled their own environmental impact statement for the Secretary. This "counter impact statement" was completed by May 4, the end of the Secretary's self-imposed 45-day waiting period.

Secretary Morton was under pressure from all sides. Both the *New York Times* and the *Washington Post* urged him to decide in favor of a Canadian common corridor. Twelve Republican Senators, most on them from the middle west, where the Canadian pipeline would terminate, wrote him in support of the Canadian alternative. At the end of March, the Canadian Minister of Energy, Mines and Resources, Donald MacDonald, visited Secretary Morton and Secretary of State Rogers to talk about pipelines. According to Minister MacDonald, he made it perfectly clear that Canada would do anything she could to make a Canadian pipeline possible, including supplying inexpensive Alberta oil to the American West Coast to ease the deficit there temporarily. Minister MacDonald also pointed out that the Canadian government had completed or nearly completed extensive environmental studies of the Mackenzie River Valley route at a cost of about $15 million and offered to make them available to the United States. (The United States had previously requested information about the dangers of the tanker route but no information about a trans Canada pipeline.) On April 28, the Canadian government decided to start construction of a 570-mile all-weather highway between Fort Simpson and Inuvik, along the Mackenzie River. And on May 4, Minister MacDonald wrote Secretary Morton assuring him that the Canadians would be ready to expedite an application to build a Mackenzie River Valley pipeline if such an application were made "by the end of this year [1972]." [148]

Also on May 4, the EDF presented the Secretary with its

"counter impact statement." Its conclusions were not surprising: that the Canadian pipeline was preferable because of the ecological damage the Alaskan pipeline could cause, the need for an overland gas route anyway, and the weaknesses of the Interior Department's economic arguments for the Alaskan route. EDF called the Interior Department document "a passive document that blandly accepts at face value the fundamental premises of the oil companies." [149] The construction stipulations, much touted during the Hickel administration and somewhat modified since then, were "vague and imprecise." And alternative ways of getting the North Slope oil to market were not really considered. Economic data were inadequate and inadequately analyzed.[150]

The EDF document contained fifty-six articles. Secretary Morton received them on May 4. On May 8, they were distributed to ten Interior Department employees. The men were given two days in which to read them, digest them, and comment. On May 10, their comments were sent to Secretary Morton, who had not read the original impact statement himself, but only portions of it selected by Deputy Undersecretary Horton.

On Thursday, May 11, Alyeska publicist Robert Miller, an Alaskan newspaperman who had joined the pipeline company after a brief stint as former Governor Miller's press secretary, was working in his cubbyhole office at Connole and O'Connell, a law firm one of whose partners, Quinn O'Connell, had represented the pipeline company since 1969. Miller had his shoes off and was drinking coffee. A reporter called to tell him that the Interior Department was putting out a statement on the pipeline by Secretary Morton that afternoon. A few minutes later, another reporter called with the same information. Miller put in calls to the three Alaskan Congressional offices to see if he could find out what the Secretary was going to say. No one in Senator Stevens' office could tell him, although the Secretary had called the Senator the evening before to tell him his decision. No one in Senator Gravel's office knew anything either. But someone in Representative Begich's office promised to check around and call Miller back.

Meanwhile, Miller told O'Connell, who then contacted friends in the Interior Department. They confirmed the reporters' information but would not tell him what the Secretary had decided. Then, Representative Begich's administrative assistant called Miller to say that Secretary Morton had decided to grant the permit for the Trans Alaska Pipeline. Miller sat down to compose Alyeska's official statement of gratified pleasure. O'Connell took a taxi to the Interior Department.

"I am convinced that it is in our best national interest to avoid all further delays and uncertainties in planning the development of the Alaska North Slope reserves by having a secure pipeline located under the total jurisdiction and for the exclusive use of the United States," said Secretary Morton. Once more, the future of the pipeline was up to the courts.

Meanwhile, in Alaska, the financial pinch was growing tighter. State officials decided the only thing to do was to take the money they needed from the pocket that had it—the oil industry's.

In early March, on the eve of hearings in the State Senate on his proposal that the state own the pipeline, Governor Egan announced that the cost of the pipeline had now risen to $3.5 billion. The Governor was worried. The state's oil revenues were then calculated on a percentage (between 15.5 and 20.5) of the wellhead price of crude oil. The more the pipeline cost to build, the lower the wellhead price would be. In fact, the Governor was afraid that, even if the pipeline were built by 1976, the state would get no royalties until after 1980 because the wellhead price would be so low. In the past, the state had estimated that it would get between $128 and $234 million a year from the North Slope, but its calculations were based on a $1.5 billion pipeline completed by mid-1967. The start of these revenues would then coincide with the depletion of the $900 million nest egg, assuming state spending continued to rise 15 percent annually, no more. But, if the pipeline was going to cost more than twice the $1.5 billion, the wellhead price would sink from $2.50 (the state's estimate of the wellhead price was 20 cents higher than the Interior Department's) to 84 cents and it would take more oil for the state to make any money at all. Because the pipeline was not expected to reach full operation until 1980 (assuming construction by 1976), with a depressed wellhead price, the state could not expect any revenues until after 1980. The Governor also feared that the oil companies would hold the daily throughput down to one and a half million barrels a day or less in order to save on taxes.

There was not much support in the state legislature for state ownership of the pipeline, although many members wanted more state control over its operations. State ownership ran counter to the laissez faire attitude of most white Alaskans. So the Governor had proposed an alternative measure, which would allow the state to regulate pipeline tariffs and obtain a 20 percent interest in the facility itself.

At the hearings, the state's witnesses argued that, with a lax Interstate Commerce Commission, clever accountants, and complicated financial maneuvers, the oil companies would make money off the North Slope while the state would not. The oil companies sent a team of lobbyists, headed by SOHIO's Chairman of the Board Charles E. Spahr, to reply that the ICC was a powerful agency with stringent regulations sufficient to protect the state's interests, which was a highly questionable assumption. Spahr and the Governor met privately to discuss a compromise. What the Governor wanted was a promise from the companies that the wellhead price would not be lower than $2.50 a barrel during the first seven years of the pipeline's operation. He did not get it.

Meanwhile, the legislature began to talk about a severance tax that would not fluctuate with the wellhead price, a minimum cents-per-barrel tax. Obviously, this could work to the state's disadvantage in low-volume oilfields, which would probably shut down rather than operate unprofitably. But on the North Slope this would not be true and the state could institute some sort of sliding scale to encourage production in smaller fields. When the legislature adjourned early Sunday morning, June 18, after the longest session in its history, it had approved such a tax. The new law provided that if the wellhead price of North Slope oil dipped below $2.65 a barrel, the companies would have to start paying a tax of 35 cents a barrel on all crude produced. The new tax was set up on a sliding scale so that low-volume fields would pay far less. The Governor's original minimum figure of $2.50 had been raised fifteen cents to ensure quick payment of the Native claims. In effect, the legislature had passed on the state's share of the Native claims settlement to the oil companies. In addition to the new severance tax, the legislature passed a right-of-way leasing act, which would allow the state to regulate pipeline tariffs. The Egan administration hoped this package would guarantee the state about $300 million a year once the pipeline was in operation.

Predictably, the companies challenged the new laws in court. Meanwhile, Governor Egan ably defended his administration against charges that the new taxes would discourage development. The oil companies were going to make huge profits in Alaska, he wrote in a letter to the editors of the *Anchorage Daily News*, and he quoted ARCO president Thornton Bradshaw as saying that the companies expected "an absolute minimum" of $1 after-tax profit on every barrel of North Slope oil (this was based on a $3.20 market price on the West Coast). As the West Coast price was expected to rise to $3.75 by the late 1970s, the Governor wrote, the oil companies' profits would also rise—to $1.55 a barrel. At an annual rate of 547.5 million barrels of oil a year, the companies' annual profits after taxes would be $484.6 million, he calculated. The federal income tax on these earnings would be about $70.1 million, not the $200 million the industry claimed it would have to pay.

12

IS ALASKA WORTH EIGHT CENTS?

In the summer of 1972, once the Native land claims had been settled, massive and unprecedented land distribution began in Alaska. By midsummer, there were twelve regional Native corporations and, on July 2, ten of them got their first checks from the federal government, for $500,000 each. (The other two got theirs a couple of days later in ceremonies at Bethel and Dillingham, which were delayed so that Representative Begich might present the money in person.) Under the settlement terms, most of this money had to be passed on to the villages, which, in turn, were to incorporate once the Native enrollment was complete. One of the most important uses for this money was to help the villages in making their land selections, which were literally the key to their futures. A mistake in land selection could mean disaster, both for the old way of life and for the village's chances in the future. Secretary Morton had set aside about 100 million acres from which the villages and the regional corporations were to pick their lands. The villages came first.

In Barrow, a haphazard cluster of jerry-built houses sandwiched between a modern airstrip and the Arctic Ocean, the future stockholders in the village corporation met in the local elementary school before a huge map of the North Slope. They were told that all they knew must be on that map—where the caribou migrate, where the best places to launch whaleboats are, where the women and children gather berries in the summer, where the birds nest, where gravel is abundant, where their homesites are, where the coal seams, oil seeps, and springs are located. In short, that map must contain their total collective life experience.

However, that meeting was only part of the process of land selection by the Barrow Natives. The village had also hired a University of Alaska biologist to help the people with an overlay, which would add geological and other data, including mineral information filed by various companies in Juneau, to what the Native people already knew. This sort of information was hard for the village people to obtain by themselves. Then there would be a second map, showing the village's aspirations and priorities. The villagers would have to consider questions such as, How could the present be changed to make the future better? and What sort of land would be important to the village ten, twenty, even fifty years from 1972?

Barrow, which has a history of being progressive, innovative, combative, and reasonably unified, was better off than many villages. In some, there was strong sentiment for dividing up the money and letting each villager fend for himself. In others, there was pressure to use the money for services and capital improvements which were, or at least should have been, provided by the state and federal governments. In some areas, the villagers were overly dependent upon white consultants and attorneys. In others, the leaders were having trouble getting the true meaning of the settlement across to their people.

Barrow's combativeness expressed itself in other ways, too. In June 1972, the people of Barrow and four other North Slope villages voted to form the world's largest unit of local government, the 56.6-million-acre North Slope Borough. One of the reasons for creating it was to finance a regional high school in Barrow with a property tax on the Prudhoe Bay installations.

Fear of what the power to tax could mean sent the oil

companies into court even before the borough was formed, so that the actual voting took place in the midst of a battle over the borough's legality. The oilmen found themselves in the awkward position of opposing local government. Since Prudhoe Bay had no permanent residents, the companies could not be represented in the borough assembly. But clearly the borough would operate almost entirely on taxes levied on the companies' property. The oilmen feared the Eskimos would adopt the highest possible tax rate for the oilfield installations. The Eskimos and their lawyers argued that the oilmen's fears were groundless, that taxes must have a relationship to needs, and that to simply levy a tax and let the money accumulate would probably be illegal.

"We have to operate under the law. We are responsible people," remarked Eben Hopson, chosen the new borough's first chairman at the same election which created it (his principal opponent was Charlie Edwardsen, who ran as a write-in candidate and was badly beaten). But he added, "In order to operate this borough effectively, I am going to need some taxes." [151]

The Eskimos finally settled on a tax designed to bring in about $4.2 million annually. More than two-thirds of the assessed property on the tax rolls, amounting to $765 million, belonged to the oil companies. The companies promptly sought and got a court order removing $565 million worth of property from the rolls. The Eskimos quadrupled the tax rate.

Later, the Arctic Slope Regional Corporation, which represented the same people as the borough, contracted with Standard Oil of California to locate and develop the underground treasure for which the Barrow leadership had fought so hard in Washington. And a subsidiary corporation laid plans for a $1.3 million 42-room hotel in Barrow.

"Our goal is to be landlords and to reap the benefits of the land," said regional corporation president Upicksoun. The Arctic Eskimos were on their way. But other Native groups were not.

Bethel, a bleak fishing town near the west coast, is in the delta of the Yukon and Kuskokwim Rivers. Its vast marshes support a wealth of birds during the summer, but its year-round human residents are desperately poor. Bethel is the home of the Calista Corporation, a shaky regional organization created by the claims act. In 1973, Calista had had three presidents in four months. The first reportedly spent $1,000 on a handsome desk so that his Bethel office would really look like an office, then hired a $45,000-a-year financial manager from Las Vegas who had to be fired a short time later.

Problems like these have made it difficult for the corporation to get down to the difficult business of land selection. Calista's 17,000 stockholders live in more than fifty villages, most of them clustered along the region's great rivers, and their claims often overlap. Unlike the Arctic Slope, there is no mineral wealth to be developed, just fish and birds and the land. Although Calista would receive federal payments for eleven years under the terms of the settlement, its more distant future looked grim. What would happen once those funds had been spent?

To many white Alaskans, already bitter about the claims settlement, the creation of the North Slope Borough was just one more example of the racial polarization that they saw taking place in Alaska in the summer of 1972. Some polarization was probably inevitable, considering the virulent anti-Native feelings whites had displayed so openly during the years the claims were before Congress for settlement.

There were moments of *opéra bouffc.* Latc in June, a state health official decided that the *Tundra Times* should not be allowed to sell muktuk, raw whalemeat considered an Eskimo delicacy, at its booth at the Tanana Valley State Fair late that summer. Eating muktuk could poison people, this official said; besides, muktuk was usually stripped from whales on beaches where dogs urinated on it. Many Eskimos took the banning of muktuk sales for a subtle form of racial discrimination. "The man is obviously not the right person to evaluate the worth of Native foods," said *Tundra Times* editor Howard Rock. "He does not have the dignity to work with this assignment." [152]

Other incidents were more serious, not so much for what they were in themselves, but for what they showed was happening beneath Alaska's seemingly carefree exterior. In May, a white columnist for the *Anchorage Daily News* wrote that, in Kotzebue, where she lived, there was a growing separation between whites and Eskimos. Jane Pender, who had lived in Arctic Alaska a long time and had worked hard to reason with white Alaskans about the justice of the Native claims, wrote:

> It pains me that what I must write about this morning is not gain, but loss; not good, but evil in its most profound sense; and that in order to write about it I must make some distinctions which I find deeply repellent.
>
> What I must say is: Here is Eskimo, and over here is white and between these two, over what has been a long, hard and very difficult winter, has grown a barrier. Our town, already deeply divided between Eskimo and Eskimo, is now further divided; this time between Eskimo and white. And none of us can foresee the future.[153]

Mrs. Pender then listed several community gathering places she felt were now "off-limits to whites." Her chief mistake was that she listed only one by name, the office of RuralCAP, the local antipoverty agency, whose head was also the mayor of the town. She continued:

> If you have never thought of yourself or others primarily in terms of race rather than of individual achievement or individual personality, it is hard to know how to deal with "White man, go home." . . . We learned that now nothing counted except race. You were either in or you were out and for reasons which had nothing to do with your own self, but only with . . . something intangible in many instances. For color itself is not a distinguishing difference between Eskimo and white.
>
> Much of what has been happening here is a positive good. It is good to see Eskimos assuming control of their own destinies. It is good that the political balance in our village has shifted into the hands of the people who are in the majority here. It is good to

> hear Eskimo spoken in public meetings. It is good to see Eskimo dancing in the schools. . . .
>
> But, it is not good that an integral part of the political power shift is racial hostility and anger; and that the few whites who live here, teachers, technicians, doctors, nurses, the small handful of business people, are being made to assume the collective burden of all the ills which the Eskimos have suffered over the years as a result of contact. It is understandable that these things would be so projected, but understandable does not mean desirable.[154]

Mrs. Pender's complaint is the traditional dilemma of the white liberal when the underdog's battle is half-won and he turns upon his benefactors.

The column provoked a rash of angry letters and was reprinted in the Native weekly. Mrs. Pender and the mayor engaged in a verbal duel, which ended with him accusing her of racism and her apologizing abjectly for inferring that he was the source of most local racist rhetoric. The important point was neither the charge nor the countercharge but the heightened racial consciousness which they revealed. It was, as Mrs. Pender herself noted, both good and bad.

While the Natives grew in economic and political power, white Alaska watched nervously from behind its barricade. The director of the Naval Arctic Research Laboratory in Barrow agonized over the tension between young whites working at the laboratory and their Eskimo counterparts in the village, a tension that often erupted into fisticuffs at the Polar Bear Theater, a makeshift movie house which becomes a dance hall at midnight after the last show. An Alaskan businessman, prominent in Chamber of Commerce circles, wondered if "we are going to have to live with two Alaskas, one white and one Native," and, in the next breath, launched into stories about improvident Natives he had known and how much their lawyers would make out of the settlement, proving that his Alaska had always had a split personality. The Attorney General characterized the formation of the North Slope Borough as indicative of a growing separatist movement among some Natives, "the tendency, if unchecked, to become an independent country."

Polarization was exactly what the Field Committee had feared when it worked out its political science textbook solution to the land claims. However, political realities made the Field Committee's recommendation that the Natives should receive little land but a sizable stake in Alaska's economic future impossible in 1971. The claims settlement was, among other things, a rare exercise in Indian self-determination. The Natives had decided that they did not want the Field Committee's solution, no matter how well-intentioned.

For all its internal problems, the AFN was probably the best-organized "Indian" group that ever dealt with the federal bureaucracy and Congress. The Natives did an extraordinary job of presenting their case, lobbying for it, walking out of negotiations at the right times, and giving in at the right times. They showed political acumen and farsightedness, which promise well for their collective political and social future, if they can stick together. And they demonstrated that they had the ability to assume their proper role in the state's economic life. Alaska will never be the same again.

The Natives will never be the same again either. The land claims settlement has ended, once and for all, the possibility of their continuing to live as their ancestors lived. To succeed under the settlement, the villagers must begin to think like white men. Thus, the settlement is probably the death knell for Native culture, despite the careful attempts of the Natives to preserve it, and this is indeed a high price to pay for assimilation into the white man's world of land deeds and land development.

The destruction of the Native way of life was probably inevitable. Any settlement would have done it. And white men have been destroying it with a vengeance ever since their first contact with the villages. The 1971 land claims settlement will be a success if it can provide a bridge between the Native world and the white one, and if it can enable most of the Natives to make an adjustment to an alien culture while preserving aspects of their own.

Since the settlement, the Native leadership has undergone a series of upheavals. Predictably, the Arctic Slope people have gone off on their own course. Much of the jealousy and suspicion with which many Natives regarded one another before 1966 has once again come to the surface. The more successful and dynamic regional corporations tend to regard their less competent counterparts with scorn. Differing regional interests have become more important than those of the Native people as a whole. But most Natives who understand the settlement at all realize that it has to work for them. It is their last chance.

Without the presence of the huge Prudhoe Bay oilfield and the industry's anxiety over the Trans Alaska Pipeline, the Native claims would never have been settled as they were. The claims were settled promptly and generously because they stood in the way of white man's progress. The need for Prudhoe Bay oil, real or imagined, made the claims a national issue rather than an Alaskan one, and because of this, the Natives got better treatment from Congress than they could have expected had their case rested solely on its merits. Had Congress treated the land claims as a purely parochial matter, the Congressmen would have listened primarily to the members of the Alaska Congressional delegation and, through them, to the multiplicity of special interests they represented, among which the Natives were only one voice and a small one at that. For none of the members of the Alaskan delegation really represented the Natives. All three tried to represent "Alaska" as they saw it. But none spoke for the Natives. As national special interests, like the oil industry and the conservationists, entered the picture, the regional special interests, which had monopolized so much of the Alaskan delegation's time, became less and less important in the total picture. The United States needed both Alaskan oil and Alaskan wilderness, so it made less and less difference what the moribund Alaska Miners Association thought.

In an administration which was not noted for its enlightened domestic policies or for its handling of minority issues, matters dealing with the Alaskan Natives were handled with unusual sensitivity and wisdom. This was probably due to the influence of Leonard Garment, a personal friend and former law partner of President Nixon's, who was White House adviser on minority matters between 1969 and 1973. It was Garment's staff which negotiated a way for the administration to support a generous settlement. This support, in turn, gave the Natives immeasurable help in Congress, particularly in the House. But one must not forget that it was also in the administration's interest to settle the claims and get the Arctic oil on stream.

The settlement process was a classic Congressional confrontation between the special interests and the public interest, but with an added twist. The Natives' interest was both special and public. If the settlement was truly no more than a real estate transaction, as many Natives liked to say it was, then the Natives were just one more group with a special interest in parceling out Alaskan land. However, the settlement was more than that. It was also a belated attempt to come to grips with a moral issue—the obligation of white Americans to the descendants of the people their own ancestors dealt with so brutally in the past. Laws were made to protect the aboriginal people, and they were broken. Other laws were made without regard for their rights as human beings or the dignity of their cultures, and these laws were enforced. More through oversight than intent, there was still a possibility in Alaska to honor the old protective laws and make a new one which, however imperfectly, tried to take into account the Natives' humanity and cultural integrity as well as the price of land in Alaska, even if for the wrong reasons. It may be too much to expect Congress to be humane and honorable unless something like billions of barrels of oil are involved.

The claims settlement was a real estate transaction, though, and in 1971 Alaskan land had another value—as wilderness. Any land settlement threatened that wilderness. Even though the settlement provided for extensive additions to the national conservation systems, its effect was to make possible the development of Alaska.

The Alaskan wilderness is awesome and sobering. It is awesome because there is so much of it. It seems to go on forever—damp fern-strewn forests, cobalt-blue glaciers, crystalline lakes, snow-crested mountains, fields of lichens and Arctic cotton. It encompasses both the massive bulk of Mount McKinley and the thumb-high tundra willows of the North Slope. It is sobering because any contact with man changes it, however slightly. Rivers have been uprooted and rechanneled by the now-motionless gold dredges near Fairbanks. But a mere footpath across the tundra fields of the Alaska Range can be almost as destructive to the land. The wilderness which seems to go on forever is no match for modern man.

The claims settlement presented conservationists with a dilemma, the choice between people or land. They started out advocating the sort of settlement which would keep the Native people in a primitive and dependent state, living in inviolable wilderness at the pleasure of the Interior Department but thus protecting the land. The conservationists sought to preserve the status quo in land use with its poverty, its injustice, and its trauma, an attempt which was no more fair than Governor Miller's attempt to preserve the political status quo. Like the state, the conservationists had to temper their demands. And, like the state, they drove a hard bargain at the end.

However, Congress made it clear in the 1971 claims act that it believed in the value of Alaskan land as wilderness. It set aside 80 million acres for conservation purposes, provided the Secretary of the Interior wanted them. On September 13, 1972, following an out-of-court settlement of the land disputes between the United States and Alaska, the Secretary set aside 78 million acres. Conservationists attacked this compromise as a back room deal although settling such a dispute was one of the roles Congress had intended for the Land Use Planning Commission. The out-of-court settlement to which conservationists objected involved principally two mineral areas—the south slope of Mount McKinley and an area on the south slope of the Brooks Range which conservationists hoped would be included in the southern part of the Gates of the Arctic park. Secretary Morton then had until December 1973 to decide which land he would recommend Congress reserve for future parks and refuges. As the Planning Commission held hearings on this next step in Alaska's development, conservation groups undertook a campaign to arouse public interest in the process, using the slogan, "Is Alaska Worth 8 Cents?"—the cost of a stamp on a letter to the Secretary.

The commission's membership was largely Alaskan and its first recommendations, presented to the Secretary in August, reflected its composition. The emphasis was upon "multiple use," that is, the opening up of as much land as possible to mineral extraction and other development. The report dealt with 78 million acres of land. The commission urged the Secretary to set aside 14 million of them for waterfowl habitats, 4 million for fish and wildlife habitats, 22.4 million for limited recreational use, and 379,000 for scientific study. But the commission also recommended that some mineral exploration and extraction be allowed on more than 60 million acres, generally the so-called less destructive type of activity such as drilling for oil and gas. And the commission wanted to make sure that resources like the vast copper reserves in the central Brooks Range would be available for use.

The commission was not unanimous in these recommendations, however. Two of its members, Richard A. Cooley, a land use expert and ardent conservationist, and Celia M. Hunter, long active in the Alaska Conservation Society, often found themselves at odds with their colleagues. In the case of the Central Brooks Range, the majority of the commission wanted an area roughly 170 miles wide and 180 miles long to be open to mineral development because of "the national and international significance of the mineral resources" there (they meant copper primarily).[155]

> We interpret the national interest to require the prudent development of metalliferous minerals, which are one of the foundation stones of our high standard of living and national security. In the absence of such development, it is very likely that the United States will be compelled to place increasing dependence upon foreign sources to satisfy domestic mineral requirements. The present unstable political situation in many parts of the world and the nation's balance of payments problems indicate to us that such reliance is an unacceptable alternative.[156]

Furthermore, "any mines which are developed will appear insignificant when viewed in the vast expanse of the Central Brooks Range. . . ." [157] It was the rationale for the Alaska pipeline all over again.

Nonsense, said Cooley and Hunter. Mineral extraction was not compatible with wilderness. Mining was always destructive, particularly open pit mining such as is used in extracting copper. Furthermore, the minerals would have to get to market, and this would mean a network of transportation corridors through the wilderness. As for the argument based on national security and the balance of payments, they observed,

> We cannot accept the concept of resource self-sufficiency in the name of "national interest" or "national security", a pursuit which with absolute certainty propels us with ever increasing speed towards total dependency on the rest of the world as we exhaust domestic resources.[158]

At the same time, an Alaska Task Force within the Interior Department was working up a set of proposals for the 78 million acres the Secretary had reserved. Like the Planning Commission, this panel was under pressure from the extractive interests, the conservationists, and the state of Alaska, all of which wanted the land for themselves. But the task force also had to contend with government agencies, which either wanted to increase their holdings in the state, like the Forest Service, or retain them, like the BLM. In the end, Secretary Morton paid more attention to his own task force than he did to the Land Use Planning Commission, to the consternation of many Alaskans, who saw in his actions a first step toward implementing that old conservationist threat to "lock up" the state in one vast preserve. On the other hand, conservationists, while rejoicing in the protection of areas like the Central Brooks Range, looked askance at a number of the Secretary's proposals, particularly the creation of three new national forests, all to be operated in accordance with the Forest Service's multiple use policies. In other words, mining and lumbering and other extractive practices would be permitted in them.

On December 18, Secretary Morton asked Congress to add 63.8 million acres to the national park and refuge systems. He recommended the creation of three entirely new national parks: one at the Gates of the Arctic in the Brooks Range, another in the Wrangell Mountains, and a third around Lake Clark in southern Alaska. All three had been areas of dispute among the members of the Planning Commission. But the new park in the Brooks Range was to include more than eight million acres of land on which mining was to be prohibited. On this point the Secretary's proposal was even more restrictive than Cooley's and Hunter's. In the Wrangell Mountains, the court-sanctioned compromise of September 1972 had already eliminated most of the mineral lands from the proposed park and many of them were now to be incorporated into the new Wrangell Mountains National Forest, where mining would be allowed. At Lake Clark, the Secretary set aside 2.6 million acres, much more than the Planning Commission had recommended, despite the protests of Cooley and Hunter. Secretary Morton also planned to enlarge Mount McKinley National Park to include parcels of land to both the north and the south. The former had been particularly coveted by the state for its minerals.

Secretary Morton also proposed nine new or expanded wildlife refuges, comprising 31.5 million acres in all. While some did not include land which conservationists considered vital, this still represented an extensive commitment to wildlife preservation in Alaska. However, the Yukon Flats, an important area for wildlife, had been split into both wildlife refuge and national forest units to allow mining in some portions. Conservationists wondered why it could not all be made a wildlife refuge.

Finally, the Secretary recommended the creation of three new national forests and the expansion of the old Chugach forest near the Wrangell Mountains, involving a total of 18.8 million acres. Outside of the Wrangell Mountains, these lands stretched along the rivers of east central and central Alaska, the Yukon, the Kuskokwim, and the Porcupine.

Nothing Secretary Morton could have done would have made everyone happy. It remains to Congress to whittle out the precise configurations of these Alaskan parks, refuges, and forests during the next five years. In that process, the conservationists, the state of Alaska, and the extractive interests will all have another go at Alaskan land. But, because of an ambiguous rider to the Alaskan Native Land Claims Act, added hastily in conference and reinterpreted later on, these parks and refuges will probably exist some day.

13
THE PROCEDURAL MINUET

"NEPA's effect is to establish a procedural minuet. . . . If you miss a step these environmentalists take the government to court."

Representative Craig Hosmer,
August 2, 1973

On August 15, 1972, Judge George L. Hart, Jr., who admitted later that he was "tired" of the case, dissolved the temporary injunction against the Trans Alaska Pipeline. The Interior Department had complied within reason with the requirements of NEPA, and the permits and rights-of-way the Secretary planned to issue were authorized by the 1920 Mineral Leasing Act, the judge ruled. His decision surprised no one.

A full complement of oilmen, attorneys, and Interior Department officials turned up in Judge Hart's courtroom to hear the arguments. So did the environmentalists. Even Dave Brower, the white-haired, pink-cheeked knight errant of the movement, stopped by to listen for a few minutes and chat with his lawyers. His own Friends of the Earth was one of the plaintiffs.

The conservationists' attorneys argued that the Interior Department had not adequately considered the total transportation system that would be involved if the Alaska pipeline were built. They were concerned about how the 26 trillion cubic feet of natural gas at Prudhoe Bay would get to market as well as the petroleum since it was obvious that the oil companies would eventually want to market it. Judge Hart seemed worried about the gas, too. He had been told the gas could neither be reinjected into the ground indefinitely nor flared (burning gas at the well is prohibited by Alaskan law).

"I am still completely uncertain as to what in the name of heaven you do with all this gas," the judge remarked to Herbert Pittle of the Justice Department. Pittle deferred to a lawyer for Alyeska, Paul Mickey, who insisted on the doubtful proposition that the gas could be reinjected forever if necessary. To reinject or not to reinject was purely a "business decision," said Mickey.

To prove the Interior Department had really considered the possibility of a Canadian gas pipeline, Mickey cited a three-page section in the impact statement which discussed the presence of the natural gas, the possibility of reinjecting it, and alternative ways of getting it to market.

"Actually the gas production and transportation seems to be treated," murmured Judge Hart, thumbing through his copy of the first volume. "They seem to call attention to the possibilities that you suggest . . ."

His tone was doubtful. This section noted, "With increasing oil production and throughput, the amount of gas would become too much for reinjection and marketing would become necessary." [159]

"Is it your feeling that it is properly disclosed beginning there . . . ?" the judge asked Mickey.

"Yes, your honor."

As for the Mineral Leasing Act, Judge Hart was skeptical that anyone could construct a modern pipeline within the fifty-four feet allowed by that law. You couldn't even turn a team of horses around in that space, he told Dennis Flannery, one of the lawyers for the conservationists. How could Congress have meant to limit pipeline rights-of-way to a space of twenty-five feet on each side, even in 1920?

The court adjourned at noon so the judge could have lunch and that afternoon he read his brief ruling to assembled lawyers, reporters, and spectators. "The final decision rests with the Supreme Court," the judge added wearily. He had seen several of his rulings reversed by the Appeals Court recently and probably expected the same fate for this one. Flannery asked that the Appeals Court be allowed to hear the case on the existing briefs, as a means of expediting it. The higher court agreed to do so and set a date in early October for a hearing *en banc,* that is, before all nine judges, an unusual move also designed to speed the case on its way.

Seven weeks later, seven of the nine appellate judges for the District of Columbia court filed into their austere marble-trimmed courtroom. One had disqualified himself for unstated reasons and a second was absent but intended to listen to the arguments on tape. The judges sat down. Everyone else sat down. The case was announced as having to do with Alaska Airlines, a regional carrier then facing bankruptcy. There was a startled silence and then Flannery recovered sufficiently to say, "Your Honor, while this case deals with everything under the sun, it does not deal with Alaska Airlines." The hearing began.

The arguments were the same ones made before Judge Hart earlier, but the conservationists' presentation was more polished and the government's less skillful. The star performer was Edward Berlin, a fluent orator who was a partner in the Washington, D.C. firm which represented the Canadian environmentalists. Several judges were so interested in what Berlin had to say that they gave him extra time in which to say it.

When the conservationists finished, the pipeline advocates had their turn. The Justice Department was represented by Edmund B. Clark, who told the judges, by way of introduction, that he had grown up amidst oilfields. (Pittle, who under Justice Department rules could not handle the case on the appellate level, watched from the sidelines.) As one lawyer-spectator remarked afterward, apropos of the large number of law students in the audience that morning, "This was not a good lesson in how to be an advocate." "This case is not made as clearly as I would like it to be," apologized Clark when asked how thoroughly the Interior Department had investigated the Canadian common corridor. "I don't think the economy has much to do with it," he said of the Secretary's decision to grant a construction permit. "After all, it's the oil companies' money." And he insisted that the possibility of a Canadian gas pipeline did not exist, although one of the consortia then studying it had told the Federal Power Commission only three weeks earlier that it would file an application to build such a pipeline by mid-1973.

Alyeska's lawyers relied heavily upon the existence of the undated memorandum by former Deputy Undersecretary Horton, which listed the relative merits of the total Alaskan system as opposed to those of the total Canadian system as proof that the Department had fully considered the Canadian common corridor. The Horton memorandum was not part of the impact statement, having been written after its release.

"Why isn't the Horton document in the impact statement?" Chief Judge David L. Bazelon wanted to know. NEPA, the judge continued, was intended "not only for the Secretary but for the information of Congress and the public" as well.

"It is the sort of question that should have been laid out in the impact statement for the benefit of the public and Congress . . . ," added Judge Harold Leventhal, "and you have it only in the hands of the Secretary."

Mickey replied that all the facts in the Horton memorandum had been in the impact statement and that Horton had merely pulled them together for the Secretary's convenience.

The judges also questioned the Department's interpretation of the Mineral Leasing Act and its use of Special Land Use Permits to grant more than fifty-four feet of right-of-way to the pipeline company.

"It seems to me that this statute is as clear as a statute can be," said Judge J. Skelly Wright.

The pipeline companies' theory had always been "take whatever you need to use," replied Robert Jordan, Alyeska's specialist on rights-of-way.

At the end of the hearing, the conservationists made no rebuttal. They did not need to. The comments of several of the judges had made it clear that they were not impressed with the defendants' arguments.

However, the court did not act quickly. After four months of deliberation, seven judges ruled unanimously on February 9, 1973 that the rights-of-way and Special Land Use Permits which Secretary Morton proposed to issue for the pipeline violated the mineral leasing law. "At the heart . . . is the following very simple point," said the court. "Congress, by enacting Section 28 [of the Mineral Leasing Act], allowed pipeline companies to use a certain amount of land to construct their pipelines. These companies have now come into court, accompanied by the executive agency authorized to administer the statute, and have said, 'This is not enough land, give us more.' We have no more power to grant their request, of course, than we have the power to increase congressional appropriations to needy recipients . . ."

The court ruled that Alyeska must go to Congress if it wanted a right-of-way of more than fifty-four feet. "Congress intended to maintain control over pipeline rights-of-way and to force the industry to come back to Congress if the amount of land granted was insufficient for its purposes," wrote Judge Wright in the majority opinion. "Whether this restriction made sense then, or now, is not the business of the courts. And whether the width limitation should be discarded, enlarged, or placed in the discretion of an administrative agency is a matter for Congress, not for this court."

The court did not rule on the NEPA issues, adding that amending the Mineral Leasing Act might take so long that the 1972 impact statement would be outdated or developments in Canada would make the Canadian common corridor arguments moot.

The court had sidestepped the central issue, but the conservationists were elated. Despite the Appeals Court's carefully written decision, the Justice Department appealed to the Supreme Court and, in an unusual move, asked it to expedite the case so construction could start as soon as possible.

Meanwhile, Congressional proponents of the pipeline introduced bills to give Alyeska its enlarged right-of-way. The Alaskan Senators proposed that Congress authorize the pipeline on grounds of national interest, thus bypassing NEPA entirely. Midwestern members of Congress, whose constituents presumably would benefit if the North Slope oil and gas went to Chicago instead of the West Coast, introduced bills authorizing construction of a gas pipeline through northeastern Alaska to the Canadian border, presumably to connect with a Trans Canada gas pipeline. Representative Udall proposed an 18-month study of all alternatives by an independent group, after which Congress would have to authorize a way of getting the Alaskan oil to market. This bill also required the Secretary of the Interior to start immediate negotiations about a Canadian pipeline with Canadian authorities. And Senator Jackson proposed giving the Interior Secretary discretion over the size of rights-of-way as part of a bill which contained common carrier and transportation corridor regulations to which the petroleum industry objected.

The Supreme Court did expedite consideration of the government's appeal, though not exactly as the defendants had planned. On April 7, the court refused to review the lower court's decision. Two days later, Secretary Morton sent a five-page letter to all members of Congress urging immediate action on simple right-of-way legislation. Citing the discrepancy between the country's energy needs and its domestic supply and some balance of payments problems arising from the need to purchase 1.7 billion barrels of foreign oil in 1972, the Secretary urged Congress not to "force a delay of this project while further consideration is given a pipeline through Canada." The following day, the Secretary and the President met at the California White House. Two weeks later, on April 18, when the President sent Congress his 1973 energy message, a list of proposals to beef up the domestic supply of energy, one suggestion was prompt construction of the Alaskan pipeline.

By April, the 1973 energy crisis was acute enough in some places, although it was more a matter of distribution than of actual shortages. One of the places where fuel was in particularly short supply was the middle west. Shortages of fuel oil had closed midwestern schools the previous winter and shortages of gasoline threatened summer harvests. While these shortages did not last long and were largely the result of monopolistic practices within the oil industry itself, they frightened Congressmen and their constituents. So, at the same time the administration was banking on the immediate need for oil to enhance the Alaska pipeline's chances in Congress, some midwestern Congressmen were determined to have a pipeline go through Canada to their own districts. The latter allied themselves with the conservationists, who realized that energy needs made the development of the North Slope reserves inevitable and had decided to back the Canadian route as the lesser of two evils. The oil industry reactivated the coalition which had successfully pushed the land claims legislation in the House in 1971—but minus the Natives. Officially, the AFN endorsed the Alaska pipeline, but the Natives were too busy at home to play much of a role. However, some of their friends, like Representative Meeds, did. And many other actors were identical to the ones in 1971—the Alaskan delegation (although Representative Don Young, a Fort Yukon entrepreneur, had replaced Representative Nick Begich, who disappeared in 1972 while flying in a small plane from Anchorage to Juneau), Senator Jackson, maritime union president Paul Hall, other members of organized labor, and Alyeska lobbyist Bill Foster.

In early March, Senator Jackson's Interior Committee held

hearings on general rights-of-way legislation, but not bills which dealt only with the Alaska pipeline. Senator Jackson's bill gave the Secretary the right to determine how large a right-of-way ought to be. It covered not only pipelines for oil but also for gas and slurry (in this case, a mixture of coal and water), canals, highways, power lines, and other means of transportation. It also created transportation and utility corridors across public lands, requiring all rights-of-way to be confined to specific areas for environmental reasons. And it specified that all pipelines authorized by the legislation be common carriers, that is, that their owners be required to transport or purchase without discrimination against any individual or company, regardless of pipeline ownership, oil and gas produced on federal lands "and other lands in the vicinity of the pipeline." What Senator Jackson, a publicly cautious proponent of the Alaska pipeline, hoped to do was amend the mineral leasing laws to give Alyeska its enlarged right-of-way while also throwing a sop to environmentalists in the form of transportation corridors and another sop, the common carrier provision, to the growing number of people who were beginning to question the anticompetitive nature of Alyeska's proposal.

Five years of hassling over ecology had obscured this aspect of the pipeline. In 1973, Exxon, Atlantic Richfield, and the British Petroleum–Standard Oil of Ohio combination controlled about 95 percent of the leased land in the Prudhoe Bay oilfield and more than 80 percent of the proposed pipeline by which that oil would move to market. Since Exxon and ARCO then also controlled much of the domestic oil production on the West Coast, the position of these three companies, once the Alaskan pipeline was in operation, would be similar to that of the state of Texas in the days when it controlled the price of domestic oil by prorationing. In fact, the antitrust division of the Justice Department had been asking questions about the companies' plans for Alaska since 1969.

These plans were designed to give the companies a stranglehold over one-fourth, and possibly more, of the United States' proven oil reserves. Estimating proven oil reserves is a little like quoting the Scriptures: anyone can find an estimate to support his argument. And this was exactly what had happened with the Alaska pipeline. First, the oil companies underestimated the North Slope's potential in order to downgrade the true value of their holdings and to deemphasize the fact that the proposed pipeline was only a first step in developing the American Arctic. But, faced with the need to convince Congress that the Alaskan Arctic was the answer to America's short-term energy needs, including those of the fuel-short middle west, pipeline proponents began to talk about there being enough oil in Alaska to justify construction of both the Alaska pipeline and one through Canada.

"Proven reserves" are a curious and conservative guess at how much oil can be extracted from a given area. In the case of Prudhoe Bay, the figure generally used by the petroleum industry was the 9.6 billion barrels estimated by DeGolyer and MacNaughton in 1968. Although ARCO started talking about 16 billion barrels some three years after that, the American Petroleum Institute stuck with the 9.6 billion figure. Estimates for the entire North Slope varied too. DeGolyer and MacNaughton had set that figure at 25 billion barrels. Their guess was corroborated by information the Navy released in 1973 about the results of its exploratory drilling in Pet Four. Prodded by Congress, which was considering legislation to open that reserve to exploration and development, the Navy estimated that there were probably between 14 and 15 billion barrels there. The U.S. Geological Survey estimate for Pet Four was higher—between 10 and 33 billion barrels.

Whichever way you counted reserves, there was a lot of oil under the North Slope. The total "proven reserves" of the United States were only 38 billion barrels in 1971. Using the most conservative estimate, 9.6 billion, the North Slope contained about one-fourth of the country's crude oil reserves. As Senator Jackson remarked, "We are really talking about a junior Persian Gulf."

In 1973, the major owners of the proposed pipeline also controlled most of the leased land at Prudhoe Bay. BP owned or controlled 55 percent of the leases covering the formation; ARCO controlled 20 percent, as did Exxon. The remaining 5 percent was held by other companies, including Mobil, Phillips, Union Oil, and Amerada-Hess, all of which owned small portions of the Alyeska venture. Most of the BP-ARCO-Exxon leases had been acquired at bargain prices prior to the discovery of oil in 1968. (The 1969 lease sale brought the state almost ten times what it had made from all previous twenty-two lease sales, including the three on the North Slope.) After 1969, other companies entered the picture. Some were major integrated companies like Texaco, Shell, and Gulf (which holds its leases jointly with BP). Others were the so-called independents of the industry, companies not involved in all four phases of the business—production, transportation, refining, and marketing. Some independents, like Colorado Oil & Gas, operated in combination with another company. Others, like H. L. Hunt, wisely associated themselves with a company which owned part of the pipeline. Such association was necessary if a company was to ship its oil out of the Arctic because of the arrangement under which the Alyeska Pipeline Service Company operated. Furthermore, ownership in the pipeline company could be expected to have an impact on future lease sales in northern Alaska, since no company would buy a lease without first giving thought to how the oil would get to market.

In addition to holding most of the leases, as far back as 1964, the three major North Slope companies had had a variety of operating agreements which further solidified their control over the resource. The 1964 Arctic Slope Agreement between Arco and Humble (now Exxon) provided for equal sharing of all exploration and development north of the Brooks Range and east of Pet Four. It also stipulated that neither company would sell its interests without first offering them to the other company. In 1968, BP became the beneficiary of this agreement, when ARCO acquired the Sinclair Oil Company, with which BP held a substantial number of joint leases.

Exxon, ARCO, and BP also negotiated a variety of unit operating agreements to allow the Prudhoe Bay field to be developed efficiently and with a minimum of waste. Unitization is not unusual. It involves pooling equipment and prorating profits and dry hole losses among the participants on the basis of individual acreage and reserves. In view of the cost of exploration and development in the Arctic, unitization made sense for the companies. It was also a means of conserving the resource. However, as a former antitrust lawyer for the Justice Department observed in a letter to Senator Floyd Haskell, the Colorado Democrat who had ousted Senator Allot in 1972 and taken a place on the Interior Committee, "the situation is entirely different when unit agreements between these major companies could cover one-fourth of all United States proven reserves."

Like most United States pipeline companies, Alyeska was owned by oil companies. Both ARCO and BP-SOHIO owned 28.1 percent; Exxon owned 25.5 percent; and the rest was divided between Mobil (8.7 percent), Phillips (3.3 percent), Union Oil (3.3 percent), and Amerada-Hess, (3 percent). The pipeline, with its capacity of 2 million barrels a day, was designed with the estimated 9.6 billion barrels of oil in mind, although its capacity could be increased considerably by a process called looping which turns a single pipeline into a pipeline system. BP controlled 5.3 billion barrels of the proven reserves at Prudhoe Bay; Exxon and ARCO each controlled 1.9 billion. The remaining reserves were distributed among the other companies roughly in proportion to their holdings in the pipeline company.

Under common industry practice, the owner oil companies control access to a pipeline. A June 1973 preliminary staff report on the Federal Trade Commission's investigation of the petroleum industry, which led to antitrust action against the eight biggest U.S. oil companies a month later, describes the way this works:

> Through the pipeline system, crude oil is transported more or less on a constant flow-pressure basis. Trunk stations can pump-in a batch of crude only when there is a slow in the flow for it and then line pressure must be increased or decreased to adjust for the desired flow speed. The scheduling of pipeline input is very

> complex and must be worked out in advance of the shipment. Because of this process, an independent crude producer may have great difficulty in securing a place in the flow, especially if he does not have storage tanks at the trunkline station and ships a relatively small amount of crude.[160]

The FTC concluded that these arrangements constituted a barrier to competition in two ways: nonowners had to pay excessive pipeline charges and they could be excluded from or limited in their use of the pipeline, despite provisions of the Hepburn Act of 1906, which declared oil pipelines to be common carriers subject to the Interstate Commerce Commission's regulation. However, Congress gave the ICC less power over oil pipelines than over railroads or gas pipelines. In fact, the ICC has worked with the API to set rates and has, in effect, regulated the regulatory process.[161] Although excessive rates pose nothing more than a bookkeeping problem for the vertically integrated owners of pipelines (they simply transfer funds from their refining operations to their pipeline operations), the independent producers and refiners actually pay the transportation charges. Furthermore, pipeline owners may employ a variety of harassing or delaying tactics to cut off or limit use of their pipeline by independents, including irregular shipping dates and limited storage facilities. It is, as one Congressman observed, "As if General Motors owned the Interstate Highway System and charged a special toll for all cars that it did not manufacture." [162]

Pipeline abuses are not new. They were among the antitrust violations which led to the dissolution of the Standard Oil colossus in 1911. However, the individual companies which rose from its ashes went right on using control of pipelines as a means of limiting competition. In fact, pipeline abuses were among the violations that formed the basis for the federal government's earlier omnibus "Mother Hubbard" antitrust suit against all the integrated oil companies and the American Petroleum Institute, which was dropped in 1940 so it would not interfere with the industry's war effort. Although portions of this case were eventually resolved, the government decided that as originally conceived it could not be tried (there were more than a hundred defendants) and pipeline abuses continued unabated.

In addition, the Alyeska pipeline was to be what is known in the industry as an "undivided interest line." In effect, the pipeline was to be seven separate pipelines, one carried on the books of each of the companies involved. Each company had a pipeline subsidiary, which was to operate a pipeline capable of a throughput equivalent to the percentage of its ownership in the Alyeska line. Thus, BP-SOHIO would own a "pipeline" across Alaska capable of transporting 562,000 barrels of crude daily. So would ARCO. Exxon's "pipeline" would be able to move 510,000 barrels. The other four "pipelines" would be much smaller. Of course, these "pipelines" would exist only on paper, but the arrangement meant that an independent shipper would have to use the facilities of one of the owners or, should his shipment be in excess of that owner's "pipeline" capacity or more than that particular owner was willing to handle, he would have to approach the other owners, a cumbersome and time-consuming process.

In late 1969, following revelation of the details of the various North Slope operating agreements as a result of litigation between the partners in TAPS, the Justice Department began a preliminary investigation into the manner in which TAPS was organized. By February 1970, the investigation was far enough along so that the Department informed Secretary Hickel, who was about to approve construction of the haul road to the North Slope, that TAPS had potential antitrust problems. In the summer of 1970, TAPS became Alyeska. The following year, the Justice Department issued a series of Civil Investigative Demands, that is, administrative subpoenas, which were approved by Assistant Attorney General Richard W. McLaren's office. However, in August 1971, Attorney General John N. Mitchell shut down the investigation, reportedly with the cryptic comment, "In view of what is going on, this is not the time." After Mitchell's veto, the investigation slid into limbo, although it was alluded to in testimony before the Joint Economic Committee of Congress in January 1972 and later that year before the House Select Committee on Small Business.

Congressional consideration of the Alaska pipeline coincided with a growing concern about the anticompetitive nature of the petroleum industry and its role in the short-term energy crisis the country was then experiencing. Senator Philip A. Hart, D-Mich., whose Antitrust and Monopoly Subcommittee had long been investigating this problem, had asked the FTC to look into it. This request eventually led to the July 17, 1973 complaint the FTC filed against Exxon, ARCO, Mobil, Texaco, Gulf, Standard Oil of California, Standard of Indiana, and Shell, charging them with having combined or agreed to engage in anticompetitive practices to monopolize the refining of crude oil into petroleum products and to maintain a noncompetitive market structure in several parts of the United States. Senator Hart and others believed the best way to deal with the problem was to prevent an oil company from engaging in more than one of the four aspects of the petroleum business. Thus there was a new undercurrent in the whole Trans Alaska Pipeline controversy.

This was immediately evident during public hearings on legislation to amend the mineral leasing laws and expedite the Alaska pipeline and in the mark-up sessions which followed. Senators Metcalf and Haskell challenged a Deputy Undersecretary about common carriers, a subject on which Jared Carter was ill prepared to testify. "But freshman senators don't say things like that," remarked an oil lobbyist on learning that Senator Haskell wanted to force the oil companies to divest themselves of Alyeska.

The pipeline bill reported out by the Senate Interior Committee in June authorized the pipeline by giving the Secretary discretion over the size of rights-of-way. Its antitrust provisions were innocuous. Like Senator Jackson's original bill, it paid lip service to the common carrier concept but only as far as oil and gas from federal lands were concerned. The oil companies had succeeded in persuading the Committee to delete "and other lands . . . ," a clause which would have further opened up the pipeline to general usage. The bill also established right-of-way corridors. And it proclaimed as Congressional policy that early delivery of North Slope oil and gas to market were in the national interest.

Senator Jackson and his colleagues on the Committee were forced to decide whether an Alaskan pipeline was preferable to a Canadian one, since that was now the central issue. After three months, they opted for the Alaskan route.

> Regardless of whether the 1969 decision of the owner companies in favor of an all-Alaska route was the wisest or the most consistent with the national interest at that time, and regardless of whether the Administration's early commitment in favor of that route was made on the basis of adequate information and analysis, *the Committee determined that the Trans-Alaska pipeline is now clearly preferable, because it could be on stream two to six years earlier than a comparable overland pipeline across Canada.* [The emphasis is the Committee's.][163]

The Committee was under pressure from environmentalists and midwesterners to authorize a nine-month study of both routes, allowing Congress to choose between them at a later date. But it concluded, "It is . . . doubtful whether further study could contribute to the accuracy of such speculations." [164]

The Committee also rejected a proposal to bypass NEPA and end the judicial hassle by declaring that Congress found the existing environmental impact statement to be in compliance with that law. This proposal was supported by Committee members Hansen and Fannin, both fast friends of the petroleum industry, and from the sidelines by the energetic Alaska Senators, both of whom had somewhat shortsightedly given up their places on the Committee once the land claims were settled. When the Committee rejected this plan, Senator Gravel announced that he would offer it on the floor as an amendment to the Committee bill.

Senator Jackson was in an awkward position. As the Senator from Washington, where much of the Alaskan oil would be refined if the Alyeska project was built, and as a prophet of the forthcoming energy crunch, he was in favor of the line's construction. But he was also the "author" of NEPA. He solved his dilemma by sitting on a fence and pointing in several directions at once. He said he did not like the way the Interior Department had handled the pipeline project to date. But he also said he was in favor of its speedy construction because the United States needed the oil as soon as possible. He promised that if construction did not begin in the spring of 1974, he would introduce legislation to break the deadlock, presumably some sort of NEPA waiver. But he publicly opposed bypassing NEPA in mid-1973, warning, "I am afraid . . . if we once establish such a precedent, that each Member of the Congress will ask and expect a legislative waiver of NEPA on his own special projects . . ." [165] In reality, Senator Jackson did little to resist the combined efforts of the Alaskans, the oil companies, organized labor, and the administration to get the pipeline authorized by name.

By the time the Senate took up the bill on July 9, Foster and the other oil lobbyists, aided by the looming energy crisis, had succeeded in subtly altering the controversy. No longer was the question *should* the pipeline be built; it had become *where* should the pipeline be built, a different issue altogether. There were to be two major amendments offered on the floor. One was Senator Gravel's amendment, designed to take the matter out of the hands of the courts and authorize the Alaskan route. The other, proposed by Senators Walter Mondale, D-Minn., and Birch Bayh, D-Ind., took it out of the hands of the Interior Department. The Mondale-Bayh amendment authorized a National Academy of Sciences study of alternative routes, to be followed by a Congressional determination of which was preferable. The State Department was directed to start discussions immediately with the Canadians about a Trans Canada pipeline. This amendment was supported by the three environmental organizations whose lawsuit had dropped the pipeline in Congress' lap in the first place. It was, in a way, also a waiver of NEPA, although its advocates did not look at it as such.

The oil companies were very clever in casting the issue as which pipeline was preferable rather than should a pipeline be built at all. Once this was done, the choice was likely to be the Alaskan pipeline since it was hard to make the more distant Canadian route sound like the answer to the United States' pressing fuel needs.

In addition to amendments dealing with the route, other important amendments reflected the antimonopoly sentiment in Congress. These ranged from Senator Haskell's proposal that the seven owner companies divest themselves of Alyeska to a sweeping attack on the vertical integration of the petroleum industry put forward by Senator Frank Moss, D.-Utah. In the end, the Senate deferred action on all of the amendments, but they probably would not have passed, had they come up for a vote.

Other amendments sought to place some restrictions upon the oil companies, for example, by limiting crude oil sales to Japan. The oilmen's strategy, wisely laid out by lobbyist Foster, was not to fight such amendments but to concentrate on defeating Mondale's and Bayh's amendment and passing Gravel's.

In mid-July, the Senate was in double session, considering one bill in the morning and another in the afternoon, in order to get through as much legislation as possible before the August recess. The pipeline debate droned on for more than a week. The floor was frequently empty except for the Senator who was speaking.

The first crucial vote came on Friday, the thirteenth, when the Senate defeated the conservationist-backed Mondale-Bayh amendment by more than two to one.[166] Voting for an independent study were eastern liberals, principally from New England, midwestern Senators whose motives varied, a handful of westerners like Metcalf and Haskell, and some individuals like Democrats Lawton Chiles of Florida and William J. Fulbright of Arkansas. All but five were Democrats. The amendment was defeated by southerners,

Republicans, and westerners, a coalition reminiscent of the one which fought off the conservationists' amendments to the land claims settlement.

A few days later it became evident that, in its zeal to defeat the Mondale-Bayh amendment, the administration had misrepresented its discussions with Canadian officials about a Canadian oil pipeline. On June 8, at the request of Representative John Melcher, Chairman of the House Public Lands Subcommittee, which was handling the pipeline legislation, the State Department had directed its embassy in Ottawa to submit a variety of questions to the Canadian government. The embassy then sent the State Department several documents indicating that the Canadian government would and could expedite construction of an oil pipeline and contradicting some of the arguments of the oil companies and the administration in favor of the Alaskan line. However, the Department's reply to Representative Melcher on June 22 was quite different in tone from the embassy's response. It included these observations: "Negotiation by the United States of a pipeline agreement with Canada does not appear possible at this time and the Departments of State and Interior are convinced that there is no Canadian alternative to the proposed Alaskan pipeline at this time," and "The Canadian government has no strong current interest in the construction of a Mackenzie Valley oil pipeline." [167] On July 6, Canadian Energy Minister MacDonald implied on the floor of the House of Commons that the State Department's letter to Representative Melcher was inaccurate. As a result, the State Department released what purported to be the questions asked the Canadian officials by the U.S. embassy and the Canadians' answers. Minister MacDonald was quoted as saying in Commons on May 22, 1973, apropos of the proposed gas (not oil) pipeline, that the Canadians would require 51 percent Canadian ownership, a condition already accepted by the consortium planning to build it. Two days after the defeat of the Mondale-Bayh amendment, the State Department included another paragraph from the original:

> In connection with an oil pipeline which might take U.S. oil to U.S. markets using Canada as a "land bridge", it would not be its policy to require majority Canadian ownership. However, as Canadian oil becomes available it would be expected that the pipeline would be expanded to accept such oil.[168]

It was a case of the State Department "acting like the front office of Exxon," remarked Senator Mondale,[169] but he admitted that having the correct information probably would not have changed the outcome of the vote.

At noon on Tuesday, July 17, the Senate voted to bypass NEPA and, to borrow a phrase from an ARCO advertisement in the *Washington Post* that morning, "get on with it." It was a dramatic moment. Forty-nine senators voted for the Gravel amendment; forty-eight against it. Since Senator John C. Stennis, D-Miss., was still recuperating from being shot during a holdup the previous winter and Senator Warren Magnuson, D-Wash., was traveling in China, ninety-eight senators should have voted. Only ninety-seven had. The unaccounted-for senator was Alan Cranston, D-Calif., who was en route to the floor when the vote was tallied. There followed some quick parliamentary manuevering by opponents of the amendment, during which Senator Cranston arrived. Then the Senate voted a second time, technically on the motion "to lay on the table," ordinarily a perfunctory move that is uncontested. This time the Senate split evenly, forty-nine to forty-nine. Vice-President Agnew, who was presiding, broke the tie. It was the first time he had done so since August 1969. The Gravel amendment passed.

Right up until the very last moment, the Alaskans and the oil lobbyists had been unsure of victory. Counting heads at a party at Senator Stevens' house the night before, they thought they still did not have enough votes. But there were some surprises, one being the vote cast by Senator Edward W. Brooke, R-Mass., one of the five Republicans to vote for the Mondale-Bayh amendment earlier. On July 17, he voted with the Nixon Administration to bypass NEPA and authorize the Alaska pipeline.

On the same day, the House Public Lands Subcommittee approved a similar waiver of NEPA, along with changes in the mineral leasing laws regarding rights-of-way for oil and gas pipelines only, and sent this bill to the full Committee. The legislation reflected the lobbyists' attentions to the subcommittee's chairman, John Melcher, a veterinarian and a former mayor of Forsyth, Montana. At first, the oil lobbyists, led by Foster, stayed away from Representative Melcher, fearing they might prejudice him against them. However, when they did finally visit him, they discovered that the Congressman was delighted to have their help drafting a bill. He got it.

When the full Committee met the following Monday, its members voted twenty to eighteen to retain the NEPA waiver. However, some members were absent and the vote was so close that conservationists were hopeful they could reverse it the following day. Representative Burton, D-Calif., switched to the winning side in order to be able to move that the vote be reconsidered. On Tuesday, these efforts also failed and the Committee approved legislation not unlike that which the Senate had already adopted. It was July 24, and Congress was expected to recess on August 3, but the bill's supporters were confident that it would get to the floor before vacation. Their confidence was not misplaced. The Rules Committee immediately gave it a rule and the leadership scheduled one day of debate, August 2. The leadership was under great pressure from oil lobbyists and the administration to act quickly. The lobbyists, as one described it, "put a blowtorch" to Speaker Carl Albert, representative of the oil state of Oklahoma and, therefore, not unattentive to their wishes. The President sent a letter to the Hill pleading that "unique" circumstances made it necessary to bypass NEPA this one time.

On Thursday, August 2, Representative Melcher took the floor to explain the rights-of-way bill, including the controversial Title II, which said Congress was satisfied with the environmental impact statement the Interior Department had produced. Proponents of the pipeline argued that this did not constitute an exemption to NEPA, merely Congressional certification of a job well done. "NEPA," said Representative Hosmer, in a particularly memorable speech,

> is a procedural arrangement whereby . . . the government is required to carefully examine environmental considerations and environmental alternatives. However, after having examined all these things NEPA does not require the government to pay a bit of attention to them. I repeat, NEPA requires only that the procedure be gone through, it does not require the government to do anything more. . . . Thus, NEPA's effect is to establish a procedural minuet. You have to go through a dance and dance this minuet with regard to environmental considerations and alternatives in the way NEPA prescribes. If you miss a step these environmentalists take the government to court. When these people go to court in these NEPA cases what they do is to say that the department or agency did not dance this minuet correctly; they did not shuffle all of the papers in the correct order or did not consider all of the alternatives or comply with some other prerequisite. It is clearly a procedural thing; it is not a substantive thing. This is because when it is all done, the department or agency is under no compulsion to give the NEPA report any weight at all.[170]

The crucial vote was on an amendment offered by Representative John Dellenback, R-Ore., which would have replaced the controversial part of Title II with language directing the courts to expedite the pipeline case, thus resolving the legal hassle without setting a precedent of bypassing NEPA, or so conservationists hoped. In support of the Dellenback amendment, Representative Udall observed that Congress was feeling the effects of "environmental backlash." "This action [Title II] . . . would have been laughed out of the chamber a year ago," he said. "Tonight a lot of those who helped to write the National Environmental Policy Act into law are preparing to gut it. . . . The issue here is whether we are going to give due process to the environment." [171]

Many members bridled at being described as "succumbing" to

the oil companies. "Well, the oil companies are trying to panic this country and panic this House and this Congress," retorted Representative Udall.[172]

The showdown came about five o'clock in the afternoon when the House rejected the Dellenback amendment, though by only twenty-three votes. Once again, liberals and eastern Congressmen found themselves aligned with some midwesterners against representatives of the south and west and Republicans. The debate continued for nearly five more hours, ranging over a wide spectrum of subjects—divesture, liability for oil spills, tanker routes, and restrictions on the sale of oil to Japan—but the companies had already won and the final vote, 356 to 60, was truly no more than a procedural minuet.

When the conferees assembled in the fall to reconcile the relatively minor differences between the House and Senate bills, there were new obstacles in the pipeline's way and also new urgency that it be built. The estimated cost of the pipeline had gone up another $1 billion, to $4.5 billion, and it seemed less and less likely that it could be completed by 1977 as Alyeska had promised. Because of the ballooning cost, which severely altered the economics of the operation, and because of continuing delays, some of the members of the Alyeska consortium were growing disenchanted with the project. In October, when Alyeska managers asked the member companies to authorize $28 million to prepare for construction the following spring, they were only able to raise $5 million. ARCO and BP-SOHIO, owners of most of the oil and gas and themselves crude-short companies, chipped in. The other five, including Exxon, refused. Exxon, with access to plenty of petroleum even in 1973, had always felt ambivalent about the pipeline and the potential competition of a BP-SOHIO combination with ready domestic crude. The four smaller companies, which owned far less of the North Slope oil, were frightened that there might not be enough profit in the venture to make their increased investment worthwhile. There was talk of selling shares in the pipeline, presumably to BP-SOHIO, which needed the line most.

In the meantime, opposition had risen to a previously little-noticed section of the Senate bill, one which strengthened the FTC's power to combat unfair and deceptive practices and allowed independent agencies to seek data from businesses without the approval of the Office of Management and the Budget (OMB). These sections had been added on the floor following desultory debate. After examining them, Roy Ash, head of OMB, decided that he opposed them. Before coming to the OMB, Ash had been head of Litton Industries and it was precisely the sort of information about Litton that he had fought giving the FTC earlier which the Senate bill would now make available. Big businesses, like General Motors, also objected to provisions which would allow the FTC to seek preliminary injunctions as a means of stopping deceptive practices and allow the FTC to go to court itself if the Justice Department failed to act within ten days. When the conference agreed in October on a bill which retained these FTC provisions, Ash and the U.S. Chamber of Commerce combined forces to lobby for recommittal to conference. The oil companies, as big businesses, were faced with a dilemma. BP-SOHIO stood behind the final bill, FTC provisions and all, but ARCO and Exxon decided not to fight recommittal. The administration was also split. OMB wanted the President to veto the bill if it reached his desk with the onerous FTC provisions intact.

However, the legislation did still contain these sections when the President signed it into law on November 16. In the meantime, the Arab countries had cut off oil supplies to the United States in an attempt to change U.S. policy toward Israel, and the pressure for opening up new and certain sources of petroleum had become irresistible. The country faced possibly severe fuel shortages during the next few years and the pipeline bill was the first of the President's "energy" measures to be approved by Congress. (Actually, the pipeline could do nothing whatsoever to relieve the short-term "crisis.") The President signed. And the oil companies won their great victory. They could now go ahead with the pipeline.

In 1968 and 1969, the companies had approached the costly venture as they clearly were accustomed to approach any such project, with supreme arrogance and callous disregard for both the public interest and the law. In fact, the companies were so disdainful that even an industry-oriented bureaucracy and a pro-pipeline Secretary of the Interior could not give them the necessary permits.

All this occurred before NEPA became effective in January 1970. Before this, nothing delayed the pipeline's construction except the usual BLM pipeline permit process, Secretary Hickel's concern for public opinion, and the companies' arrogant stupidity. After January 1970, the environmentalists had a new weapon. Using NEPA, lawyers for three conservation groups were able to prove that the Interior Department viewed complying with the law as a sort of game, Representative Hosmer's "minuet." The only question was how complex the minuet had to be to satisfy the courts. The Department started out with an eight-page impact statement which grew in two years to a nine-volume document. And the companies were forced to consider and incorporate environmental safeguards they clearly never intended.

What was never considered, however, neither in the Interior Department, nor in the courts, nor in the Congress, was whether the pipeline should have been built at all. In view of the growing demand for more oil, it was probably inevitable, but, to those concerned about the environment, it set a sad precedent.

On December 3, only a few days after the President had signed the Federal Land Rights-of-Way bill expediting construction of the Alaska pipeline, William E. Simon, a tough-minded New York investment banker, was sworn in as the country's first energy czar. Simon said on that occasion, "We must increase our domestic supply of energy. . . . We can no longer delay development of our domestic energy resources. For example, we will push for the development of Naval Petroleum Reserve #4, the building of a second Alaska pipeline. . . ." [173]

At the same time, a report was circulating in the Defense Department which proposed construction of a $4 to $5 billion pipeline from Pet Four to an unspecified ocean terminal on Alaska's southwestern coast. And this was not the second Alaska pipeline to which Simon referred on December 3.

14

FALL 1974

By late summer of 1974, construction had begun on the Trans Alaska Pipeline. It had been five years since the Twentieth Science Conference met at Fairbanks to discuss the future of Alaska. Many things had happened in those five years and yet the state was, curiously, no more ready for the actual advent of the pipeline than it had been in 1969. And the pipeline's cost had risen to $5.9 billion.

By fall 1974, the chief topic of conversation in Alaska was neither the windfall wealth the state (and presumably its people, too) would enjoy from the massive project nor the merits of the conservationists' arguments against it. It was the impact the pipeline's construction would have on the average Alaskan. Nowhere would the impact be greater than in Valdez, the terminal of the pipeline.

Valdez is a small town of low, ranch-style buildings, spread out along the shore of the Valdez Arm of Prince William Sound, a deep fjord in which enormous tankers will wait to load up with Prudhoe Bay oil, collected at a tank farm across the bay. Old Valdez, a picturesque little Gold Rush harbor nestled on a river delta at the foot of spectacular fog-enshrouded mountains, was washed away during the 1964 earthquake. In five minutes on March 27, a thirty-foot tidal wave sank a coastal supply freighter that was unloading Easter lilies at the Valdez dock, swept through the little town destroying everything in its way, and then receded, leaving thirty-one people dead and wreckage where the packing plant, dock, cannery, local hotel and bar, and other familiar buildings had been. Later, the rubble was bulldozed under and new Valdez built on geologically sounder foundations.

In new Valdez, Alaskans were coming face to face with a trauma equal to the Easter Earthquake, as Dr. Victor Fischer of the state university's Institute of Social, Economic, and Government Research had predicted they would. Alyeska planned to bring in 3,500 men and to house them in a work camp that would be more than three times the population of the town. Tiny Valdez was expected to swell even further—to about 8,000—with non-pipeline workers and their families who had come to Alaska to make their fortune in the modern Gold Rush. There was inadequate housing. In the summer of 1974, people were camping wherever they could on the outskirts of the town. The schools foresaw an onslaught of students even if Alyeska provided them with modular buildings to relieve the congestion. Supplies had always been scarce and expensive in Valdez. But prices were expected to soar at the town's two grocery stores and many other items would simply continue to be unavailable. A family would still have to drive to Anchorage, more than 300 miles away, for some basic needs. And the residents of this idyllic fjord were worrying about crime, particularly rape. Who could blame an Alaskan who wondered, a little guiltily given the widespread faith in bonanza, if the Trans Alaska Pipeline was worth it?

Elsewhere in Alaska, the struggle for the land continued. The Natives were making their choices; the state had already made its. And the federal government hovered over everything, nearly as omnipotent as it was in territorial days and, for many Alaskans, still as remote. Conservationists hoped to enlarge upon the Interior Department's plans for more than 80 million acres of parklands, refuges, national forests, and other special use preserves. They were opposed by mining interests, by the state, and in many instances by the Natives. The ultimate outcome will be the result of negotiations between the parties and Congressional action during the coming years.

The issues were familiar ones—the development of mineral deposits, the value of land as wilderness, the wise definition of "multiple use" in regard to public lands, the clash of Native interests with those of white men. For example, there was the question of roads.

For years, the state of Alaska has had an ambitious plan for highways throughout this virtually roadless state. Many are physically and economically improbable, like the road to Nome which legislators from that distant town on the coast of the Bering Sea have been proposing at almost every session of the legislature since statehood. Some proposed roads are opposed by the residents (usually Native) of the towns they are intended to "open up"; some by Natives through whose land the roads would pass. Many would bisect areas conservationists want to preserve as wilderness or, at least, relatively inaccessible parkland. Some, of course, would help the people in the towns they would connect with civilization and are favored by them.

These things were the same. Others were different. In late August, former Governor Hickel lost his attempt at a political comeback to a white big game guide, fisherman, and poet from the Native village of Naknek on Bristol Bay. Jay Hammond won the Republican gubernatorial nomination (and later the governorship) on a promise of controlled growth during the period of the pipeline boom. It was a long way from the 1967 promise to get Alaska moving again economically which won former Governor Hickel his first public office.

And to the surprise of many Alaskans, homesteading in the state came to an end in May 1974 when Public Land Order 5418 was published. Thus ended a great tradition of the Great Land. Homesteading in Alaska was seldom successful and never easy but the wilderness is dotted with tiny cabins. The smoke curls up from the chimney and there is a sign on a tree warning hunters of the presence of children and pets. This is possibly the last real alternative to twentieth century America. Its demise is certainly the beginning of the end of the last frontier.

NOTES

1. Summer 1969

1. "Conservationists, Oilmen Clash at Conference," *Anchorage Daily News*, August 26, 1969, p.1.
2. Ibid.
3. Rogers, George W., *Change in Alaska: People, Petroleum and Politics*, University of Alaska Press, College, Alaska, 1970, p.48.
4. Ibid., p.164.
5. Ibid., p.172.
6. Ibid., p.134.
7. Ibid., p.71.
8. Ibid., p.118.
9. Ibid.
10. "Stevens Blasts Conservationists," *Anchorage Daily News*, August 28, 1969, p.1.
11. "Stevens Tees Off—Again," *Anchorage Daily News*, August 29, 1969, p.1.
12. "Stevens Blasts Conservationists."
13. "Stevens Tees Off—Again."
14. "Stevens Blasts Conservationists."
15. Rogers, pp.202–203.
16. Pender, Jane, "Only One Native . . . ," *Anchorage Daily News*, September 4, 1969.
17. Ibid.

2. Two Alaskas

18. Federal Field Committee for Development Planning in Alaska, *Alaska Natives & The Land*, Anchorage, Alaska, October, 1969, pp.236–38.
19. See Chance, Norman A., *The Eskimo of North America*, Holt, Rinehart & Winston, New York, 1966.
20. Federal Field Committee, p.431.
21. Ibid., p.432.

22. Potter, Jean C., *Alaska Under Arms*, Macmillan Company, New York, 1942, p.161.

23. Hearings on Alaska Statehood, Committee on Interior and Insular Affairs, U.S. Senate, 81st Congress, Second Session, April 24–29, 1950, p.293.

24. Hearings on Statehood for Alaska, Subcommittee on Territories and Insular Possessions, U.S. House of Representatives, 83rd Congress, First Session, April 14–17, 1953, p.85.

25. Hearings on the nomination of Governor Walter J. Hickel of Alaska to be Secretary of the Interior, Committee on Interior and Insular Affairs, U.S. Senate, 91st Congress, First Session, January 15, 16, 17, 18, 20, 1969, p.16.

26. Rogers, George W., *The Future of Alaska: Economic Consequences of Statehood* (sponsored by the Arctic Institute of North America and Resources for the Future, Inc.), Johns Hopkins University Press, Baltimore, 1962, p.148.

27. Hearings on Alaska Statehood and Elective Governorship, Committee on Interior and Insular Affairs, U.S. Senate, 83rd Congress, First Session, August 17, 18, 19, 20, 24, 25, 1953, pp.60–61.

28. Hearings on Alaska Statehood, Committee on Interior and Insular Affairs, U.S. Senate, 83rd Congress, Second Session, January 20–29, February 1–4, 24, 1954, p.227.

29. Ibid., p.229.

30. U.S., *Statutes at Large*, 72 Stat. 339, Sect. 4.

3. The Natives Awake

31. Editorial, *Tundra Times*, Fairbanks, Alaska, March 18, 1963.

32. Federal Field Committee, p.13.

33. Editorial, *Tundra Times*, March 18, 1963.

34. Editorial, *Tundra Times*, October 1, 1962.

35. "Senator Proposes Cash Payment for Valid Native Land Claims," *Tundra Times*, April 15, 1966, p.1.

36. Hensley, Willie, letter to the editor, *Tundra Times*, April 22, 1966.

37. "Gruening Clarifies Position on Native Land Claims," *Tundra Times*, April 29, 1966, p.1.

38. Editorial, *Tundra Times*, September 23, 1963.

39. Hopson, Eben, letter to the editor, *Tundra Times*, July 22, 1963.

40. Editorial, *Tundra Times*, July 20, 1964.

41. Nicholls, Hugh, letter to the editor, *Tundra Times*, June 24, 1966.

42. "Statewide Native Unity Achieved at Conference," *Tundra Times*, October 28, 1966, p.1.

43. *Anchorage Daily News*, December 22, 1966, p.2.

44. Ibid., November 22, 1966, p.2.

4. Native Against White

45. Hearings on A Bill to Authorize the Secretary of the Interior to grant certain lands to Alaska Natives, Settle Alaska Native Land Claims, and for other purposes, Committee on Interior and Insular Affairs, U.S. Senate, 90th Congress, Second Session, February 8, 9, 10, 1968, p.134.

46. Ibid., p.172.

47. Ibid., pp.275–76.

48. Ibid., p.278.

49. Ibid., p.171.

50. Ibid., p.50.

51. Ibid., p.285.

52. Ibid., p.303.

53. Ibid., p.382.

54. Ibid., p.189.

55. Ibid.

56. Ibid., p.265.

57. Ibid., p.237.

58. "Justice Goldberg Quits AFN," *Tundra Times*, May 16, 1969, p.1.

59. Ibid.

60. Notti, Emil, "Position with respect to the Natives land claims issue," Alaska Federation of Natives memorandum, June 20, 1969.

61. Hearings on Alaska Native Land Claims Settlement, Committee on Interior and Insular Affairs, Subcommittee on Indian Affairs, U.S. House of Representatives, 91st Congress, First Session, August 4, 5, 6, September 9, 1969, p.138.

62. Hearings on a bill to provide for the settlement of certain land claims of Alaska Natives, Committee on Interior and Insular Affairs, U.S. Senate, 91st Congress, First Session, April 29, 1969, p.154.

63. "The Goldberg Bill," editorial, *Anchorage Daily Times*, October 18, 1969.

64. Ibid.

65. "Proper Commentary," editorial, *Anchorage Daily Times*, October 21, 1969.

66. Ibid.

67. "Unity on Land Claims," editorial, *Anchorage Daily Times*, October 22, 1969.

68. Hearings on Alaska Statehood, U.S. Senate, April 24–29, 1950, pp.293–94 (see n.23).

69. Atwood, Robert B., "Statehood Is Achieving What Its Advocates Wanted," *Anchorage Daily Times*, January 3, 1969, p.1.

70. Miller, Governor Keith H., letter to Senator Henry M. Jackson, November 18, 1969.

71. Ibid.

72. "Claims Opponents are Scolded by Stevens," *Anchorage Daily News*, November 25, 1969, p.1.

73. "The Prevailing Mood," editorial, *Anchorage Daily Times*, November 25, 1969.

74. "Meanwhile, Sen. Stevens Hammers Away," *Anchorage Daily News*, November 26, 1969, p.1.

75. "General Agreement Among Native Leaders with Notti," *Tundra Times*, February 13, 1970, p.1.

76. Ibid.

77. Hickel, Secretary of the Interior Walter J., letter to Senator Henry M. Jackson, February 18, 1970.

78. "Council Admits State Claims Role," *Anchorage Daily Times*, January 10, 1970, p.3.

79. Ibid.

80. Ibid.

5. Oil Comes To Alaska

81. Hearings on Alaskan Submerged Lands, Committee on Interior and Insular Affairs, U.S. Senate, 85th Congress, Second Session, April 25, May 14, 22, 1958, p.19.

82. Editorial, *Fairbanks News-Miner*, Feb. 2, 1968.

83. "Word Is Get Going Fairbanks," *Fairbanks News-Miner*, August 10, 1968, p.1.

84. "Aggressive Planning Needed Now," editorial, *Fairbanks News-Miner*, August 12, 1968.

85. "City Leaders Visit Oil Executives," *Fairbanks News-Miner*, December 7, 1968, p.1.

86. "Brookings Seminar Tackles the Challenge that Alaska Is," *Anchorage Daily News*, November 11, 1969, p.1.

6. Private Industry's Most Expensive Undertaking

87. Hughes, George, letter to Bureau of Land Management, Anchorage and Fairbanks, June 6, 1969.

88. Train, Undersecretary of the Interior Russell E., letter to R.E. Dulaney, June 10, 1969.

89. Dulaney, R.E., letter to Undersecretary of the Interior Russell E. Train, June 19, 1969.

90. U.S. Department of the Interior, internal memorandum, June 20, 1969.

91. U.S. Department of the Interior, internal memorandum, June 18, 1969.

92. Ibid.

93. Ibid.

94. Hickel, Secretary of the Interior Walter J., letter to R.E. Dulaney, June 27, 1969.

95. Memorandum to the President of the United States, "Preliminary

Report of the Federal Task Force on Alaskan Oil Development," September 15, 1969.

96. Hickel, Secretary of the Interior Walter J., letter to Senator Henry M. Jackson, December 3, 1969.

97. U.S. Department of the Interior, *Environmental Impact Statement on Haul Road Adjacent to Trans-Alaska Pipeline*, March 20, 1970.

98. "Join to Push Pipeline," *Fairbanks News-Miner*, April 13, 1970, p.1.

99. Pecora, Dr. William T., memorandum to Secretary of the Interior Walter J. Hickel, March 25, 1970.

100. "Hickel Assures Eventual Permit," *Fairbanks News-Miner*, April 23, 197O, p.1.

7. A Land Settlement

101. "AFN Pres. 'Optimistic' About Claims Bill in Congress," *Anchorage Daily News*, December 13, 197O, p. 1.

8. The Million-Acre Kisses

102. Parmeter, Adrian, Alaska Federation of Natives memorandum for the U.S. Department of the Interior, January 18, 1971.

103. Wright, Don, letter to Secretary of the Interior Rogers C.B. Morton, February 24, 1971.

9. Justice Or Oil?

104. Brandborg, Stewart M., editorial, *The Living Wilderness*, Spring 1971, p.2.

105. Hearings on Bills to provide for the settlement of certain land claims of Alaska Natives, Committee on Interior and Insular Affairs, U.S. Senate, 92nd Congress, First Session, April 29, 1971, p.489.

106. Ibid., p.509.

107. Hearings on bills to provide for the settlement of certain land claims of Alaska Natives, Subcommittee on Indian Affairs of the Committee on Interior and Insular Affairs, U.S. House of Representatives, 92nd Congress, First Session, May 3–7, 1971, p.276.

108. Ibid., p.300.

109. Ibid., p.314.

110. "Natives Grab the Ball," editorial, *Fairbanks News-Miner*, April 7, 1971.

111. Bass, Kenneth, internal memorandum for the Alaska Federation of Natives, September 7, 1971.

112. "Billion Dollar U.S. Settlement Raises Canadian Native Hopes," *Toronto Star*, September 30, 1972.

113. Congressional Record, October 19, 1971, p.HR0733–9734.

114. Ibid., p.HR9741.
115. Ibid.
116. Ibid., October 20, 1971, p.HR9796.
117. Ibid., p.HR9801.
118. Ibid., p.HR9805.

10. The Delicate Compromise

119. *Congressional Record*, November 1, 1971, p.S17287.
120. Ibid.
121. Ibid., p.S17313.
122. "Natives Hit Egan on Claims Stand," *Anchorage Daily News*, November 12, 1971, p.1.
123. Wright, Don, letter to members of conference on Alaska Native Land Claims bills, November 16, 1971.
124. McGovern, Senator George, letter to members of conference on Alaska Native Land Claims bills, November 17, 1971.
125. Wright, Don, letter to members of conference on Alaska Native Land Claims bills, December 1, 1971.
126. *Congressional Record*, December 13, 1971, p.HR12355.

11. On With the Pipeline

127. U.S. Department of the Interior, *Draft Environmental Impact Statement for the Trans-Alaska Pipeline*, January 1971, pp.152–54.
128. Ibid., p.167.
129. "Arctic Slope Natives Adamant," *Anchorage Daily Times*, January 12, 1971, p.1, and "Alaskans Hail Federal Report Approving Arctic Oil Pipeline," *New York Times*, January 15, 1971, p.11.
130. "Pipeline Realities," editorial, *Fairbanks News-Miner*, March 2, 1971.
131. "Friday, March 5: Day Bad News Rained on Alaska," *Fairbanks News-Miner*, March 6, 1971, p.1.
132. Figures released in April 1971 by the State Department of Revenue.
133. "Oil Companies Should Review Entire Alaska Pipeline Project Says Morton," *Anchorage Daily Times*, March 3, 1971, p.1.
134. "Company Not Considering Canada Route, Says Patton," *Anchorage Daily Times*, April 3, 1971, p.1.
135. Alaska Department of Economic Development, *Pacific Rim Trade Opportunities*, September 1971, p.11.
136. *Anchorage Daily Times*, June 11, 1971, p.1.
137. Saylor, Representative John P., and Udall, Representative Morris, letter to Secretary of the Interior Rogers C.B. Morton, January 11, 1972.
138. Ibid.

139. Hearing on amendments to the Alaska Native Claims Act of 1971, Committee on Interior and Insular Affairs, U.S. Senate, 92nd Congress, Second Session, March 2, 1972.

140. Cameron, Roderick A., Director and Watson, H. Lee, Staff Scientist, Environmental Defense Fund, Inc., letter to Secretary of the Interior Rogers C.B. Morton, February 22, 1972.

141. National Audubon Society, Sierra Club, Trout Unlimited, and Wilderness Society, telegram to Secretary of the Interior Rogers C.B. Morton, March 15, 1972.

142. "Oil Pipe Dreams?," *Newsweek*, April 3, 1972, p.62.

143. Tussing, Arlon; Rogers, George W.; and Fischer, Victor, *Alaska Pipeline Report*, Institute of Social, Economic, and Government Research, University of Alaska, Fairbanks, 1971, p.19.

144. Ibid., p.22.

145. Ibid., p.90.

146. U.S. Department of the Interior, *An Analysis of the Economic and Security Aspects of the Trans Alaska Pipeline*, Summary, Vol. I, December 1971, pp.E-18–E-21.

147. Alyeska Pipeline Service Company, Inc., *The Environment*, March 1972.

148. MacDonald, Donald, Minister of Energy, Mines and Resources, letter to Secretary of the Interior Rogers C.B. Morton, May 4, 1972.

149. Flannery, Dennis M.; Hillyer, Saunders S.; and Barnes, James N., letter to Secretary of the Interior Rogers C.B. Morton, May 4, 1972.

150. Dienelt, John F., and Stoel, Thomas B., Jr., letter to Secretary of the Interior Rogers C.B. Morton, May 4, 1972.

12. Is Alaska Worth Eight Cents?

151. "Eben Hopson Confident of Job—Borough Chairman Fills Job with Past Experience," *Tundra Times*, June 28, 1972, p.1.

152. "Food Discrimination?," ibid., p.2.

153. Pender, Jane, "Eskimos vs. Whites," *Anchorage Daily News*, May 28, 1972, p.35.

154. Ibid.

155. Joint Federal-State Land Use Planning Commission for Alaska, *Land Use Planning and Policy in Alaska*, Vol. I, National Interest Lands, Part I, report to the Secretary of the Interior, July 1973, p.III-5-5.

156. Ibid.

157. Ibid.

158. Ibid., p.III-5-10.

13. The Procedural Minuet

159. U.S. Department of the Interior, *Final Environmental Impact Statement on Proposed Trans Alaska Pipeline*, March 1972, Vol. I, pp.75–76.

160. *Congressional Record*, July 16, 1973, p.S13597.

161. For a discussion of this government-industry relationship see Engler, Robert, *The Politics of Oil: Private Power and Democratic Directions*, University of Chicago Press, Chicago, 1961, p.324.

162. Representative Donald Frasier, D-Minn., in *Congressional Record*, June 26, 1973, p.HR5465.

163. Report on Federal Lands Rights-of-way Act of 1973, Committee on Interior and Insular Affairs, U.S. Senate, 93rd Congress, First Session, Report #93-207, June 12, 1973, p.21.

164. Ibid., p.22.

165. *Congressional Record*, July 17, 1973, p.S13671.

166. The vote was 61 to 29 against the Mondale-Bayh amendment.

167. *Congressional Record*, July 17, 1973, p.S13671.

168. Ibid.

169. Ibid., p.S13672.

170. Ibid., August 2, 1973, p.H7238.

171. Ibid., p.H7276.

172. Ibid., p.H7277.

173. U.S. Department of the Treasury, December 3, 1973 press release, "Statement by the Hon. William E. Simon, Deputy Secretary of the Treasury, on Federal Energy Administration and Energy Policy," p.7.

BIBLIOGRAPHY

Adelman, M.A., editor, *Alaskan Oil: Costs and Supply*, Praeger, New York, 1971.

Alaska Department of Economic Development, *Pacific Rim Trade Opportunities*, September 1971.

Alyeska Pipeline Service Company, Inc., *The Environment*, March 1972.

Alyeska Pipeline Service Company, Inc., *Summary, Project Description of the Trans Alaska Pipeline System*, August 1971.

Bass, Kenneth, internal memorandum for the Alaska Federation of Natives, September 7, 1971.

Brown, Tom, *Oil on Ice*, Sierra Club, San Francisco, 1971.

Caldwell, Lynton K., *Environment: A Challenge for Modern Society*, Doubleday & Co., Anchor Books, Garden City, 1971.

Cameron, Roderick A., Director, and Watson, H. Lee, Staff Scientist, Environmental Defense Fund, letter to Secretary of the Interior Rogers C.B. Morton, February 22, 1972.

Chance, Norman A., *The Eskimo of North America*, Holt, Rinehart, & Winston, New York, 1966.

Cooley, Richard A., *Alaska: A Challenge in Conservation*, University of Wisconsin Press, Madison, 1966.

———, *Politics and Conservation: The Decline of the Alaska Salmon*, Harper & Row, New York, 1963.

———, and Wandesforde-Smith, Geoffrey, editors, *Congress and the Environment*, University of Washington Press, Seattle, 1970.

Cooper, Bryan, *Alaska—The Last Frontier*, Hutchinson & Co., London, 1972.

Dienelt, John F., and Stoel, Thomas B. Jr., letter to Secretary of the Interior Rogers C.B. Morton, May 4, 1972.

Dulaney, R.E., letter to Undersecretary of the Interior Russell E. Train, June 19, 1969.

Engler, Robert, *The Politics of Oil: Private Power and Democratic Directions*, University of Chicago Press, Chicago, 1961.

Federal Field Committee for Development Planning in Alaska, *Alaska Natives and the Land*, Anchorage, October 1968.

Federal Task Force on Alaskan Oil Development, memorandum to the President of the United States, "Preliminary Report of the Federal Task Force on Alaskan Oil Development," September 15, 1969.

Flannery, Dennis M.; Hillyer, Saunders; and Barnes, James N., letter to Secretary of the Interior Rogers C.B. Morton, May 4, 1972.

Gallagher, H.R., *Etok: A Story of Eskimo Power*, G.P. Putnam's Sons, New York, 1974.

Gruening, Ernest, *The State of Alaska*, Random House, New York, 1968.

Hearings on Alaska Statehood, Committee on Interior and Insular Affairs, U.S. Senate, 81st Congress, Second Session, April 24–29, 1950.

Hearings on Statehood for Alaska, Subcommittee on Territories and Insular Possessions, U.S. House of Representatives, 83rd Congress, First Session, April 14–17, 1953.

Hearings on Alaska Statehood and Elective Governorship, Committee on Interior and Insular Affairs, U.S. Senate, 83rd Congress, First Session, August 17–20, 24–25, 1953.

Hearings on Alaska Statehood, Committee on Interior and Insular Affairs, U.S. Senate, 83rd Congress, Second Session, January 20–29, February 1–4, 24, 1954.

Hearings on Alaskan Submerged Lands, Committee on Interior and Insular Affairs, U.S. Senate, 85th Congress, Second Session, April 25, May 14, 22, 1958.

Hearings on A Bill to Authorize the Secretary of the Interior to grant certain lands to Alaska Natives, Settle Alaska Native Land Claims, and for other purposes, Committee on Interior and Insular Affairs, U.S. Senate, 90th Congress, Second Session, February 8–10, 1968.

Hearings on the Nomination of Governor Walter J. Hickel of Alaska to be Secretary of the Interior, Committee on Interior and Insular Affairs, U.S. Senate, 91st Congress, First Session, January 15–18, 20, 1969.

Hearings on a bill to provide for the settlement of certain land claims of Alaska Natives, Committee on Interior and Insular Affairs, U.S. Senate, 91st Congress, First Session, April 29, 1969.

Hearings on Alaska Native Land Claims Settlement, Committee on Interior and Insular Affairs, Subcommittee on Indian Affairs, U.S. House of Representatives, 91st Congress, First Session, August 4–6, September 9, 1969.

Hearing on Oversight of Oil Development Activities in Alaska, Special Subcommittee on Legislative Oversight, Committee on Interior and Insular Affairs, U.S. Senate, 91st Congress, First Session, August 12, 1969.

Hearings on the Status of the Proposed Trans-Alaska Pipeline, Committee on Interior and Insular Affairs, U.S. Senate, 91st Congress, First Session, September 9, 1969.

Hearings on Bills to provide for the settlement of certain land claims of Alaska Natives, Committee on Interior and Insular Affairs, U.S. Senate, 92nd Congress, First Session, April 29, 1971.

Hearings on bills to provide for the settlement of certain land claims of Alaska Natives, Subcommittee on Indian Affairs of the Committee on Interior and Insular Affairs, U.S. House of Representatives, 92nd Congress, First Session, May 3–7, 1971.

Hearings on Oil Prices and Phase II, Subcommittee on Priorities and Economy in Government, Joint Economic Committee, 92nd Congress, First Session, January 10–12, 1972.

Hearing on amendments to the Alaska Native Claims Act of 1971, Committee on Interior and Insular Affairs, U.S. Senate, 92nd Congress, Second Session, March 2, 1972.

Hearings on Natural Gas Regulation and the Trans-Alaska Pipeline, Joint Economic Committee, 92nd Congress, Second Session, June 7–9, 22, 1972.

Hickel, Secretary of the Interior Walter J., letter to R. E. Dulaney, June 27, 1969.

———, letter to Senator Henry M. Jackson, December 3, 1969.

———, letter to Senator Henry M. Jackson, February 18, 1970.

Hughes, George, letter to Bureau of Land Management, Anchorage and Fairbanks, June 6, 1969.

Joint Federal-State Land Use Planning Commission for Alaska, *Land Use Planning and Policy in Alaska*, Vol. I, National Interest Lands, Part I, report to the Secretary of the Interior, July, 1973.

MacDonald, Donald, Minister of Energy, Mines and Resources, letter to Secretary of the Interior Rogers C.B. Morton, May 4, 1972.

McGovern, Senator George, letter to members of conference on Alaska Native Land Claims bills, November 17, 1971.

Miller, Governor Keith H., letter to Senator Henry M. Jackson, November 18, 1969.

Notti, Emil, Alaska Federation of Natives memorandum, "Position with respect to the Natives land claims issue," June 20, 1969.

Parmeter, Adrian, Alaska Federation of Natives memorandum for the U.S. Department of the Interior, January 18, 1971.

Pecora, Dr. William T., memorandum to Secretary of the Interior Walter J. Hickel, March 25, 1970.

Potter, Jean C., *Alaska Under Arms*, Macmillan Company, New York, 1942.

Report on Federal Lands Rights-of-Way Act of 1973, Committee on Interior and Insular Affairs, U.S. Senate, 93rd Congress, First Session, Report #93-207, June 12, 1973.

Rogers, George W., *Alaska in Transition: The Southeast Region*, Johns Hopkins University Press, Baltimore, 1960.
———, *Change in Alaska: People, Petroleum and Politics*, University of Alaska Press, College, 1970.
———, *The Future of Alaska: Economic Consequences of Statehood*, Johns Hopkins University Press, Baltimore, 1962.
Saylor, Representative John P., and Udall, Representative Morris, letter to Secretary of the Interior Rogers C.B. Morton, January 11, 1972.
Train, Undersecretary of the Interior Russell E., letter to R. E. Dulaney, June 10, 1969.
Tussing, Arlon; Rogers, George W.; and Fischer, Victor, *Alaska Pipeline Report*, Institute of Social, Economic and Government Research, University of Alaska, Fairbanks, 1971.
Udall, Stewart L., *The Quiet Crisis*, Holt, Rinehart, & Winston, New York, 1963.
United States Department of the Interior, *An Analysis of the Economic and Security Aspects of the Trans-Alaska Pipeline*, December 1971.
———, *Draft Environmental Impact Statement for the Trans-Alaska Pipeline*, January 1971.
———, *Environmental Impact Statement on Haul Road Adjacent to Trans-Alaska Pipeline*, March 20, 1970.
———, *Final Environmental Impact Statement, Proposed Trans-Alaska Pipeline*, March 1972.
———, internal memorandum, June 18, 1969.
———, internal memorandum, June 20, 1969.
Wilderness Society, Environmental Defense Fund, and Friends of the Earth, *Comments on the Environmental Impact Statement for the Trans-Alaska Pipeline*, May 1971.
Wright, Don, letter to Secretary of the Interior Rogers C.B. Morton, February 24, 1971.
———, letter to members of conference on Alaska Native Land Claims bills, November 16, 1971.
———, letter to members of conference on Alaska Native Land Claims bills, December 1, 1971.

Newspapers and Periodicals

Anchorage Daily News
Anchorage Daily Times
Audubon
Congressional Record
Fairbanks News-Miner
The Living Wilderness
Newsweek
New York Times
New York Times Magazine
Oil & Gas Journal
Platt's Oilgram
Toronto Star
Time
Tundra Times, Fairbanks, Alaska
Washington Post

INDEX

ARCTIC
Barrow
Wainwright
CHUKCHI SEA
U. S. S. R.
Point Hope
BROOKS
Arctic Circle
Kotzebue
KOBUK R.
SEWARD PENINSULA
KOYUKUK R.
Nome
NORTON SOUND
St. Lawrence Island
YUKON R.
BERING SEA
KUSKOKWIM R.
Bethel
Nunivak Island
Pribilof Islands
BRISTOL BAY
ALASKA PENINSULA
Aleutian Islands